Ubiquitin - Proteasome Pathway

Edited by Xianquan Zhan

Published in London, United Kingdom

IntechOpen

Supporting open minds since 2005

Ubiquitin - Proteasome Pathway
http://dx.doi.org/10.5772/intechopen.87547
Edited by Xianquan Zhan

Part of IntechOpen Book Series: Biochemistry, Volume 18
Book Series Editor: Miroslav Blumenberg

Contributors
Donald Spratt, Emma Kane, Noriyuki Murai, Gunnar Dittmar, Marta Mendes, Daniel Perez Hernandez, Marene Landström, Jie Song, Xianquan Zhan, Jianglin Zhang, Jia Guo, Miaolong Lu

First published in London, United Kingdom, 2020 by IntechOpen
IntechOpen is the global imprint of INTECHOPEN LIMITED, registered in England and Wales, registration number: 11086078, 5 Princes Gate Court, London, SW7 2QJ, United Kingdom
Printed in Croatia

British Library Cataloguing-in-Publication Data
A catalogue record for this book is available from the British Library

Additional hard and PDF copies can be obtained from orders@intechopen.com

Ubiquitin - Proteasome Pathway
Edited by Xianquan Zhan
p. cm.
Print ISBN 978-1-83880-432-9
Online ISBN 978-1-83880-841-9
eBook (PDF) ISBN 978-1-83880-842-6
ISSN 2632-0983

For EU product safety concerns:
IN TECH d.o.o., Prolaz Marije Krucifikse Kozulić 3, 51000 Rijeka, Croatia, info@intechopen.com or visit our website at intechopen.com.

IntechOpen Book Series
Biochemistry
Volume 18

Xianquan Zhan received his M.D. and Ph.D. training in preventive medicine at the West China University of Medical Sciences from 1989 to 1999. He received his post-doctoral training in oncology and cancer proteomics at the Central South University and University of Tennessee Health Science Center (UTHSC). He worked at UTHSC and Cleveland Clinic in USA from 2001 to 2012, and achieved the rank of Associate Professor at UTHSC. After that, he became a Full Professor at the Central South University and Shandong First Medical University, and an Advisor of MS/PhD graduate students and postdoctoral fellows. He is also a Fellow of the Royal Society of Medicine, Fellow of EPMA, European EPMA National Representative, full member of the American Society of Clinical Oncology (ASCO), member of the American Association for the Advancement of Sciences (AAAS), Editor-In-Chief of the International Journal of Chronic Diseases & Therapy, Associate Editor of the EPMA Journal and BMC Medical Genomics, and Guest Editor of Frontiers in Endocrinology and Mass Spectrometry Reviews. He has published more than 130 articles, 23 book chapters, 4 books, and 2 US patents in the field of clinical proteomics and biomarkers.

Editor of Volume 18:
Xianquan Zhan
University Creative Research Initiatives Center
Shandong First Medical University

Book Series Editor: Miroslav Blumenberg
NYU Langone Medical Center, New York, USA

Scope of the Series

Biochemistry, the study of chemical transformations occurring within living organisms, impacts all of life sciences, from molecular crystallography and genetics, to ecology, medicine and population biology. Biochemistry studies macromolecules - proteins, nucleic acids, carbohydrates and lipids –their building blocks, structures, functions and interactions. Much of biochemistry is devoted to enzymes, proteins that catalyze chemical reactions, enzyme structures, mechanisms of action and their roles within cells. Biochemistry also studies small signaling molecules, coenzymes, inhibitors, vitamins and hormones, which play roles in the life process. Biochemical experimentation, besides coopting the methods of classical chemistry, e.g., chromatography, adopted new techniques, e.g., X-ray diffraction, electron microscopy, NMR, radioisotopes, and developed sophisticated microbial genetic tools, e.g., auxotroph mutants and their revertants, fermentation etc. More recently, biochemistry embraced the 'big data' omics systems.

Initial biochemical studies have been exclusively analytic: dissecting, purifying and examining individual components of a biological system; in exemplary words of Efraim Racker, (1913 - 1991) "Don't waste clean thinking on dirty enzymes." Today however, biochemistry is becoming more agglomerative and comprehensive, setting out to integrate and describe fully a particular biological system. The "big data" metabolomics can define the complement of small molecules, e.g., in a soil or biofilm sample; proteomics can distinguish all the proteins comprising e.g., serum; metagenomics can identify all the genes in a complex environment e.g., bovine rumen. This Biochemistry Series will address both the current research on biomolecules, and the emerging trends with great promise.

Contents

Preface

The ubiquitin-proteasome pathway consists of ubiquitin, substrate proteins, E1 enzymes, E2 enzymes, E3 enzymes, and proteasome. Ubiquitin is a highly conserved small protein with 76 amino acids and about 8.5 kDa. E1 enzymes are ubiquitin-activating enzymes. E2 enzymes are ubiquitin-conjugating enzymes. E3 enzymes are ubiquitin-ligases. Proteasome is a 26S complex, an organelle in the cell, which contains one 20S core and two 19S lids. The ubiquitin-proteasome pathway consists of a series of enzymatic reactions: E1 binds ubiquitin to activate ubiquitin in ATP-dependent fashion, the activated ubiquitin is conjugated with E2, and then ubiquitin-conjugated E2 in concert with E3 ligases recognizes substrate proteins and chemically covalently attaches ubiquitin (monomer or polyubiquitin chain) to substrate proteins (ubiquitinated proteins). The ubiquitinated proteins are delivered to the proteasome for degradation into peptides and amino acids to be used for synthesis of new proteins. Here, substrate proteins include surplus proteins and misfolded proteins in a cell or tissue. Also, there are the deubiquitinating enzymes that can remove the attached ubiquitin chain. Thus, ubiquitination/deubiquitination is a reverse process in cells. The ubiquitin-proteasome pathway plays crucial roles in degrading most intracellular proteins, and maintaining the balance between protein synthesis and degradation. The changes of components in the ubiquitin-proteasome pathway are associated with multiple pathophysiological processes, such as cancers, and neurodegenerative diseases. For example, the mutated or overexpressed E3 ligases can act as oncogenes, and also some E3 ligases and deubiquitinating enzymes are tumor suppressors. Moreover, the ubiquitin-proteasome pathway is involved in multiple biological processes, including DNA repair, mitophagy, angiogenesis, RTK signaling, NF-kB signaling, and mitochondrial maintenance, which are dynamically regulated by ubiquitination. Also, it is involved in synaptic functions to regulate the functions of the nervous system.

This book focuses on the changes of the components of the ubiquitin-proteasome pathway, the methodology to study the ubiquitin-proteasome pathway, protein ubiquitination, and application of the ubiquitin-proteasome pathway in different diseases. Chapter 1 addresses the branching and mixing – new signals of the ubiquitin signaling system in the following aspects: the ubiquitin-conjugating system, different ubiquitin-like modifications, ubiquitin-chain topology (homotypic chains, heterotypic chains, and ubl-ubiquitin chains), and detection methods of ubiquitinated targets and chains including biochemical and genetic methods, mass spectrometry-based methods, ubiquitin topology analysis, and detection of branched chains. Chapter 2 addresses the ubiquitin-independent proteasomal degradation mediated by antizyme, which enriched the concept and content of the ubiquitin-proteasome pathway: ubiquitin-dependent proteasomal degradation through ubiquitination, and ubiquitin-independent proteasomal degradation through antizyme. Chapter 3 addresses lys63-linked polyubiquitination of transforming growth factor beta type I receptor (TBRI) specifies oncogenic signaling, and the regulation of its associated signaling pathways. Chapter 4 addresses ubiquitination and deubiquitination, and their potential clinical application value in melanoma. Chapter 5 addresses the new discoveries of more members (AREL1, HACE1, HECTD1, HECTD4, G2E3, and TRIP12) of the HECT E3 ubiquitin ligase family and their biological functions and activities except for the classical E6AP and NEDD4 family, whose dysfunction is closely associated with different diseases. Chapter 6 uses the quantitative ubiquitinomics to investigate the abnormal ubiquitination of the ubiquitin-proteasome system in lung squamous cell carcinoma, which provides important insight into understanding the nature and importance of these alterations in the ubiquitin-proteasome system in a cancerous relative to normal lungs.

This book presents the new advances in concepts, methodology, and application of the ubiquitin-proteasome pathway. However, one must realize that this book contains only a fraction of the very important ubiquitin-proteasome pathway studies in medical sciences, which serves as a spur to stimulate and encourage researchers who study the ubiquitin-proteasome pathway to come forward with its scientific merits to research and clinical practice of the ubiquitin-proteasome pathway, especially in the aspects of clarification of molecular mechanisms and discovery of new therapeutic targets and drugs for diseases.

Xianquan Zhan, MD, PhD, FRSM
Professor of Cancer Proteomics and PPPM,
University Creative Research Initiatives Center,
Shandong First Medical University,
Jinan, Shandong, China

Chapter 1

Branching and Mixing: New Signals of the Ubiquitin Signaling System

Daniel Perez-Hernandez,
Marta L. Mendes and Gunnar Dittmar

Abstract

Posttranslational modifications allow cells and organisms to adapt to their environment without the need to synthesize new proteins. The ubiquitin system is one of the most versatile modification systems as it does not only allow a simple on–off modification but, by forming a chain of ubiquitin molecules, allows conveying multiple signals. The structure of the chains is dependent on the linkage to the previous ubiquitin molecule as every lysine can serve as an acceptor point for this modification. Different chain types code for specific signals ranging from protein degradation to protein targeting different cellular compartments. Recently the code of ubiquitin signals has been further expanded as branching and mixing of different chain types has been detected. As an additional layer of complexity, modifications of the ubiquitin chain by ubiquitin-like modifiers, like NEDD8, SUMO, or ISG15, have been found. Here we will discuss the different chain types and the technical challenges which are associated with analyzing ubiquitin topology-based signaling.

Keywords: ubiquitin, chain topology, ubiquitin-like, branched ubiquitin

1. The ubiquitin-conjugating system

Since its discovery in the 1980s, the ubiquitin signaling system has gained recognition as one of the most versatile, yet complicated, posttranslational signaling systems. The central component of the system is the small protein ubiquitin (76 aa). Ubiquitin itself is always expressed as an immature precursor protein, either fused to a ribosomal protein or as a head-to-tail fusion of five or six ubiquitin moieties. The precursor protein is processed co-translationally and cleaved off the fusion protein right at the ribosome [1] liberating the mature ubiquitin protein. The modification of a target protein with ubiquitin as a PTM utilizes an enzymatic cascade. In the first step, an ubiquitin-activating enzyme (E1) is binding ubiquitin while hydrolyzing one molecule of ATP to AMP forming a thiol ester of ubiquitin's C-terminus with a cysteine residue in its active center. The activated ubiquitin can then be transferred to an ubiquitin-conjugating enzyme (E2) which again forms a thiol ester with a cysteine in its active center. Depending on the cascade which is used, the final transfer is catalyzed by one of three classes of ubiquitin ligases,

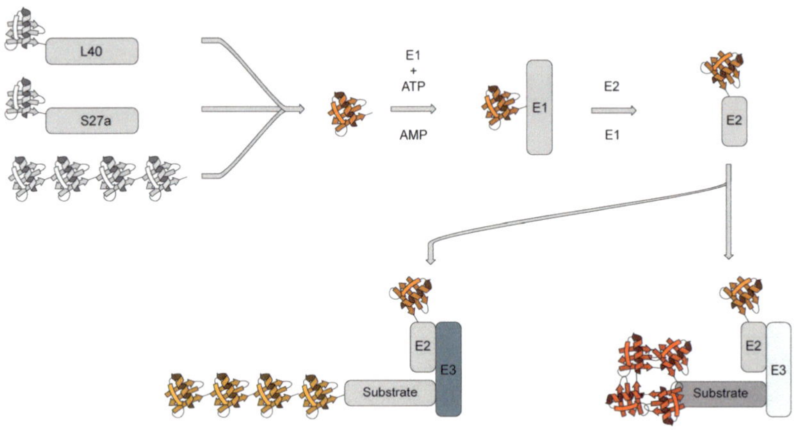

Figure 1.
The ubiquitin-conjugating system. Ubiquitin (gray folded structure) is expressed as fusions either with ribosomal subunits or as ubiquitin multimers, which are cleaved co-translational. The mature ubiquitin (orange) is released and under the consumption of one ATP bound to the activating enzyme (E1). The activated ubiquitin is then passed on to the conjugating enzyme (E2), which finally catalyzes the transfer to a substrate protein involving an E3 ligase. The reaction can then be repeated and catalyzes the formation of a polyubiquitin chain. Depending on the lysine residue in ubiquitin used, the chain can have different structures, as indicated in yellow for a linear ubiquitin chain or in red for a K48-linked chain.

really interesting new gene (RING), ring between ring (RBR), or homologous to E6 C-terminus ligases (HECT ligases) (**Figure 1**). While RING-type ligases are associating with the E2 enzyme and bringing the target protein in close proximity to the E2/E3 complex, RBR and HECT-E3 are able to bind ubiquitin itself. The final transfer of ubiquitin to the target is then catalyzed without the help of an E2 enzyme.

This modification leads to a single modification of the substrate protein with ubiquitin but can also be extended by multiple rounds of modification with ubiquitin itself being the acceptor of the modification, leading to the formation of the ubiquitin chain. The structure of the chain is dependent on which linkage is used within the ubiquitin chain. The chains highly varies in shape from a very compact structure for a K48-linked chain [2–4] to a long stretched shape in the case of K63 [5, 6] and linear ubiquitination [3]. Each of the different chain topologies is associated with different biological functions. Besides the signaling through different chain topologies, ubiquitin signaling can also occur through modifications by a single ubiquitin (monoubiquitin) or multiple monoubiquitinations.

Like for many other PTM systems, the ubiquitin signaling system has possibilities to erase the signal by either disassembling the polyubiquitin chain or by removing ubiquitin from its target. These enzymes are called ubiquitin hydrolases or ubiquitin de-conjugating enzymes (DUBs). Most DUBs belong to the enzymatic class of cysteine hydrolases, which carry a cysteine in their active center. Research on DUB specificity has shown that these enzymes possess a high linkage specificity indicating clear regulatory functions in the cell and are not acting as simple quality check enzymes [7].

2. Ubiquitin-like modifiers (Ubl)

Besides ubiquitin, there are a number of proteins which share significant similarity to ubiquitin. They fall essentially into two groups: proteins with a ubiquitin-like domain and small proteins with a similar size as ubiquitin. The small Ubls like ubiquitin, SUMO, NEDD8, Urm1, Apg8, Apg12, ISG15, Fat10, and Ufm1 are highly conserved among eukaryotes. Many of these small proteins have been found to be covalently conjugated to a substrate protein, utilizing their own conjugation cascades, which usually consists of an activating enzyme and a conjugating enzyme (see **Table 1**).

2.1 NEDD8

Neuronal precursor cell-expressed developmentally downregulated protein 8 (NEDD8) is structurally the closest relative to ubiquitin. Its E1 enzyme is split into two different parts which are forming the activation enzyme (NAE1/UBA3) [8]. The activated NEDD8 moiety is transferred to the NEDD8-conjugating enzyme E2s (UBC12/UBE2M or UBE2F) and substrate-specific NEDD8 E3 ligase [9]. Unlike the large group of ubiquitin E3 ligases, there are only about 10 different NEDD8 E3 ligases [9], and most of them belong to the group of RING E3 ligases.

The best-characterized physiological neddylation substrates are the cullin proteins (Cul-1, Cul-2, Cul-3, Cul-4A, Cul-4B, and Cul-5) which form the backbone structure of cullin-RING ligases (CRLs). The conjugation of the cullin subunit modulates the activity of E3 ligase [10]. Deconjugation of the NEDD8 is catalyzed by the signalosome complex which removes NEDD8 from the cullin [11–14].

Recently, several other proteins have been identified, which are modified by NEDD8 including ubiquitin itself, p53, mouse double minute 2, and epidermal growth factor receptor (EGFR) [15–17].

2.2 SUMO

Small ubiquitin-related modifier (SUMO) is probably the best-studied Ubl protein. It is highly conserved among eukaryotes with one gene in lower eukaryotes like baker's yeast (Smt3) [18] which developed into three homologs in humans, SUMO-1, SUMO-2, and SUMO-3. SUMO-2 and SUMO-3 are closely related, while SUMO-1 is more divergent. SUMO-1 does not form polymeric chains, while SUMO-2 and SUMO-3 mainly form K11-linked homotypic SUMO chains [19, 20]. SUMO-1 can be linked to the end of a poly-SUMO-2/SUMO-3 chain, effectively terminating chain growth [20]. Like NEDD8, SUMO is activated by a dimeric activating enzyme (SAE1/SAE2). The recognition of the target proteins is done by the conjugating enzyme Ubc9 which recognizes the main SUMOylation motif ΨKxE (Ψ = hydrophobic residue) [21–23]. Some reactions are further enhanced by the action of other E3 ligases, like RANBP2. These E3 ligases catalyze the transfer by recruiting the substrate or catalyzing the transfer of SUMO from Ubc9 [24, 25]. Similar to other Ubls, modification with SUMO can be reversed by specific proteases as summarized in Pichler [26].

Most SUMO-1 is conjugated to RANGAP1 near the nuclear pore. SUMO-2 is partially cytosolic, while SUMO-3 is mainly located in nuclear bodies. In unstressed cells, most SUMO-2 and SUMO-3 are not conjugated. Upon stress induction

Ubl	E1	E2
NEDD8	UBA3 (UBE1C), APPBP1 (NAE1)	UBE2M
SUMO	SAE1, SAE2	UBE2I
ISG15	UBA7	UbcH8

Table 1.
Ubiquitin-like modifiers that have been found linked to ubiquitin chains and their enzymatic activation cascade.

(e.g., folding stress) both SUMO-2 and SUMO-3 get conjugated to target proteins [27]. SUMO-1 conjugation has been proposed to regulate trafficking between nucleus and cytosol but also change protein–protein interaction [26, 28, 29]. SUMO modification plays an important role in a number of cellular processes like DNA replication, cell cycle regulation, nuclear trafficking, signal transduction, and protein degradation [30–32]. Recently, large-scale studies identified more than 1000 SUMO targets and increased the number of cellular processes even further [33].

2.3 ISG15

Unlike other small ubiquitin-like modifiers interferon-stimulated gene product 15 (ISG15) contains two ubiquitin-fold domains. It is massively induced by interferon treatment, ischemia, DNA damage, and aging as a monomer as well as a conjugated protein (reviewed in [34]). ISGylation requires a three-step enzymatic cascade involving an E1 activating enzyme (Ube1L), an E2 conjugating enzyme (UbcH8), and an E3 ligase (Herc5 or TRIM25/EFP) [35]. ISGylation is reversed by Ub-specific protease USP18 [36].

Several protein targets and cellular functions have been identified, which are regulated by ISGylation. These include the regulation of DNA damage response, autophagy, protein synthesis, the downregulation of the ubiquitin-proteasome system, and the regulation of HIFα in response to hypoxia [37–42].

A particular interest is the finding that modification of nascent proteins by ISGylation occurs after viral infection [43]. Virus infection induces host antiviral responses, including induction of type I interferons [44–47]. The transcription factor IRF3 recruits HERC5 and induces conjugation of ISG15 onto IRF3. This modification attenuates the interaction between Pin1 and IRF3, thus antagonizing IRF3 ubiquitination and degradation. Consistently, host antiviral responses are boosted or crippled in the presence or absence of HERC5, respectively [48–50].

3. Ubiquitin chain topology

Ubiquitination occurs in proteins at one or multiple lysine residues. Ubiquitin itself, containing seven lysine residues, can be ubiquitinated at each one of these lysine residues, as well as at the N-terminal methionine [51, 52]. Proteins can be monoubiquitinated, where a single ubiquitin is conjugated to a lysine residue in the substrate; multi-monoubiquitinated, where a single ubiquitin is conjugated to multiple lysine residues in the substrate; or polyubiquitinated, where the ubiquitin conjugated to the substrate is ubiquitinated itself. Polyubiquitinated chains can be divided into homotypic chains and heterotypic chains, and like for linear chains, different chain topologies lead to different structures and functions in the cell [53–58]. To add to this complexity, ubiquitin chains themselves can also be modified by ubiquitin-like modifiers (**Figure 2**).

3.1 Homotypic chains

Homotypic chains are composed of several ubiquitins linked together through the same lysine or N-terminal methionine residues. This leads to a total of eight possible chain types. Each chain adopts a different conformation: K6, K11, and K48 adopt "compact" conformations, while K27, K29, K33, K63, and M1 adopt "open" conformations, allowing recognition of these chains by different ubiquitin-binding partners implicated in several signaling pathways [53–59]. A short description of the functions of homotypic chains is given below.

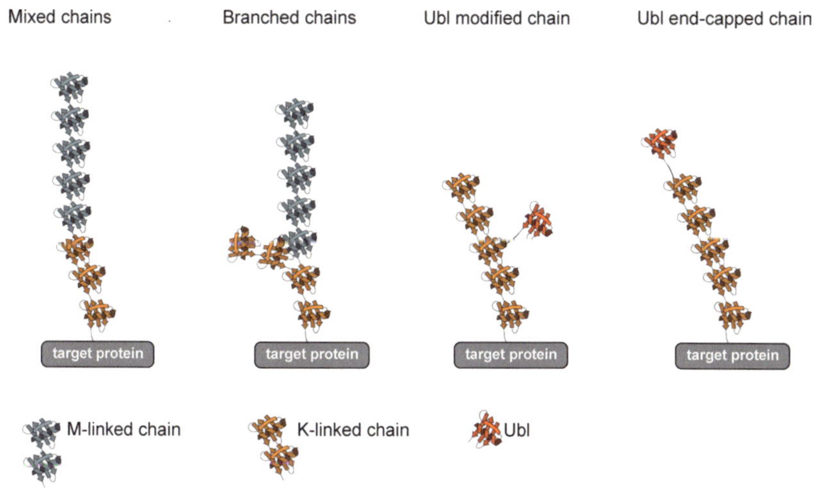

Figure 2.
Complex ubiquitin chains. Homotypic ubiquitin chains can be extended by a different type of chain leading to a mixed ubiquitin chain. If the ubiquitin chain is modified not at the last ubiquitin moiety, a branched ubiquitin chain is created. For mixed ubiquitin/Ubl chains, two attachment points are possible: either as a cap structure, modifying the last ubiquitin, or as a branching point on one of the ubiquitins in the middle.

Studying K6 chains is challenging since this chain is among the less abundant ubiquitin chains [60, 61]. Although its function is still not very clear, K6 has been implicated in mitophagy regulation [62–64] and DNA damage response [65]. More recently Michel et al. showed that HECT E3 ligase HUWE1, previously implicated in cellular processes like DNA repair, stress response, cell death, differentiation, and mitophagy [66, 67], assembles K6 chains [68]. Mitophagy is the process by which cells maintain the energy metabolism by removing damaged mitochondria. During this process, PINK1 accumulates on the surface of the mitochondrial outer membrane and recruits cytosolic PARKIN, an E3 ubiquitin ligase [69]. PARKIN then ubiquitinates mitochondrial proteins by generating canonical (K48 and K63) and noncanonical chains (K6 and K11) eventually leading to mitophagy [63]. USP30 is the only known DUB anchored to the mitochondria outer membrane which has been seen to act as a regulator of mitophagy. Despite having been seen to cleave K6, K11, K48, and K53 chains, USP30 prefers K6 chains [62, 64]. Under normal conditions, USP30 prevents mitophagy of normal mitochondria by maintaining ubiquitination at low levels. Under stress conditions and mitochondrial damage, PARKIN is recruited, highly increasing ubiquitination levels and inducing mitophagy. PARKIN and PINK1 are both mutated in patients with Parkinson's disease [70, 71].

Although not many roles are known for K11 chains, it has been shown that these chains are key players in cell cycle regulation and proteasome degradation. The anaphase-promoting complex/cyclosome (APC/C) is an E3 ubiquitin ligase and essential for cell cycle regulation. Along with Ube2C, APC/C targets key players in the cell cycle, like securin and cyclin B1, for proteasomal degradation by assembling K11 chains, thus allowing the transition from metaphase to anaphase [72–75]. In 2013, Mevissen et al. showed the ovarian tumor (OTU) DUBs Cezanne's and Cezanne2's linkage specificity towards K11 chains [76]. In 2014, Bremm and co-workers described Cezanne as a new regulator of HIF1α homeostasis [77], where HIF1α is ubiquitinated with K11 chains. The knockdown of Cezanne increases the amount of K11 polyubiquitin chains and decreases the activity of HIF1α. HIF1α degradation was not disrupted by inhibition of the proteasome suggesting an alternative degradation pathway—possible through chaperone-mediated autophagy—to HIF1α [77]. Cezanne can bind and disassemble K11 chains on APC/C substrates stabilizing them leading to cell proliferation [78]. Finally, K11 chains were shown to replace K48 chains in the transcription factor Met4 activating it [79], so far only seen to be ubiquitinated with K48 chains leading to transcription repression [80]. Although the exact composition of the newly synthesized K11 chains is still not known, the authors suggest that these chains can either be homotypic K11 chains or heterotypic K11/K48 chains [79].

K27 chains are still the least studied of all ubiquitin chains. E3 protein ligase HACE1 has been shown to assemble K27 chains onto both optineurin and YB-1 [81, 82], indicating a role in secretion through the multivesicular body (MVB) pathway. The ubiquitin ligase RNF168 assembles K27 ubiquitin chains on chromatin linking them to the DNA damage response pathway [83]. During pathogen infection, K27 chain assembly triggers immune response through the recruitment of TBK1 [84, 85]. The NEDD4 family E3 ligases, Itch and WWP2, promote K27 polyubiquitination of SHP-1 enhancing the strength of the T-cell receptor (TCR) signal and in turn negatively regulating in T_H2 cell differentiation [86]. USP19, a deubiquitinating enzyme, removes K27 chains from TRIF, thus inhibiting its recruitment by TLR3/TLR4 and consequently inhibiting TLR3−/TLR4-mediated innate immune response [87]. The E3 ligase Hectd3 assembles K27 chains on Malt1 and Stat3 promoting differentiation of pathogenic TH17 cells in experimental autoimmune encephalomyelitis (EAE), a mouse model for human multiple sclerosis [88].

K29 has been implicated in the Wnt signaling pathway. The E3 ubiquitin ligase, EDD, promotes K29 ubiquitination of β-catenin leading to higher protein levels and enhanced activity [89]. E3 ubiquitin ligase SMURF1 promotes K29 ubiquitination of axin, thus disrupting its association with LRP5/LRP6 and inhibiting the Wnt signaling pathway [90]. The ubiquitin thioesterase ZRANB1, also known as TRABID, preferentially hydrolyzes K29 and K33 chains [4, 55, 57, 91]. Although TRABID/ZRANB1 was proposed to bind and hydrolyze K63 chains from the APC tumor suppressor protein acting as a positive regulator of the Wnt signaling pathway [92], no evidence has been shown, linking TRABID and K29 chains in the Wnt signaling pathway. More recently, TRABID has been implicated in epigenetic regulation, where it regulates Il12 and Il23 genes by TLR. TRABID associates with and stabilizes Jmjd2d by hydrolyzing K29 chains regulating histone modifications and expression of Il12a, Il12b, and Il23a genes [93].

K33 chains have seen to be implicated in autoimmunity. RING-type E3 ligase Cbl-b and HECT-type E3 ligase Itch assemble K33 chains to the TCR-ζ and negatively regulate its phosphorylation and consequently TCR signaling [94]. Later in 2014, Lin et al. reported the inhibition of the type I IFN signaling due to the interaction of the DUB USP38 and TBK1, after viral infection. USP38 hydrolyzes K33 chains from TBK1 promoting K48 polyubiquitination by DTX4/TRIP for its degradation through NLRP4 [95]. During infection with uropathogenic *E.coli*, compartmentalized TLR4 signaling is triggered, and TRAF3 is K33 polyubiquitinated leading to the expulsion of intracellular bacteria by the exocyst complex [96]. After TGF-β stimulation, USP2a removes K33 polyubiquitin chains from the TGFBR promoting the recruitment of R-SMADs and consequently promoting metastasis [97]. A role of K33 chains in trafficking was also suggested. KLHL20, a BTB domain-containing adapter protein, recruits Cul-3 assembling K33 chains to Cm7. The ubiquitination of Cm7 promotes its recruitment to the trans-Golgi network (TGN) [98].

More recently, K33 chains have been implicated in autophagy. E3 ligase RNF166 binds and assembles K33 chains to the autophagy adaptor p62. This mechanism seems to be essential in targeting bacteria to autophagy [99]. SMURF1 induces K29- and K33-linked polyubiquitin of UVRAG, triggering a mechanism that promotes autophagosome maturation and inhibits HCC growth. TRABID/ZRANB1 forms a complex with UVRAG and cleaves SMURF1-induced K29- and K33-linked polyubiquitin chains from UVRAG inhibiting autophagosome flux and leading to poor prognosis [100].

M1 chains and their importance in signaling pathways and disease have been extensively reviewed [101, 102]. M1 chains are generated by the linear ubiquitin chain assembly complex (LUBAC), composed by HOIP, HOIL-1, and SHARPIN [103–105]. M1 chains play a crucial role in TNF signaling and immune response [104–109]. Formation of complex 1 starts when TNF binds TNFR1, resulting in the recruitment of TRADD and RIPK1. After binding to TNFR1, TRADD recruits TRAF2 which then recruits cIAP1 or cIAP2, two E3 protein ligases which assemble K11, K48, and K63 chains to several components of the complex. LUBAC is then recruited by the complex and assembles M1 chains to several components of the complex including NEMO, an essential modulator of NF-κB, RIPK1, TRADD, and TNFR1. This process then recruits the TAK1-TAB protein complex and the IKK complex (IKKα, IKKβ, and NEMO). While the TAK1-TAB complex activates MAPK cascades triggering JNK and p38 MAPK leading to the activation of the transcription factor AP-1, the IKK complex activates NF-κB signaling. In the absence of LUBAC, complex 2 instead of complex 1 is formed leading to cell death either by apoptosis or programmed necrosis.

K48 is the most abundant and well-studied ubiquitin chains and a major signal for proteasome degradation [60, 61, 110]. It was initially proposed that signaling through a polyubiquitin chain was mandatory for proteasome degradation [111–113]. However, several publications seem to show that monoubiquitination can also target proteins for proteasome degradation [114–116].

K63 is the second most abundant ubiquitin chain in cells [60, 61] and it is implicated in immune response, DNA repair, endocytosis, and protein trafficking (reviewed in [117–119]). K63 chains seem to be essential for the activation of the IKK signaling pathway through TRAF6 [120]. TRAF6 also induces K63 ubiquitination of Akt which is then phosphorylated, activated, and recruited to the membrane [121]. RIG-I is regulated by K63 chains where TRIM25 assembles K63 chains to RIG-I [122]. K63 ubiquitination of IRAK1 is required for the activation of NF-κB signaling [123]. INF-β signaling pathway is activated when STING is ubiquitinated with K63 chains assembled by TRIM56 [124]. K63 ubiquitination and protein trafficking and DNA damage have been extensively reviewed [119, 125]. Two of the most well-studied examples on protein trafficking are the MHC I and EGFR. MHC class I molecules are polyubiquitinated with K63 chains leading to degradation by an endolysosomal pathway [126]. EGFR is also ubiquitinated and promotes its internalization [127, 128]. More recently, it has been shown that K63 chains bind to DNA, enhancing the recruitment of repair factors [129].

3.2 Heterotypic chains

Heterotypic chains are composed of several ubiquitin molecules linked together through different lysines or N-terminal methionine residues and can be classified as mixed chains or branched chains. In mixed chains, each ubiquitin is modified only once by another ubiquitin molecule, while in branched chains each ubiquitin can be modified by two or more ubiquitin molecules. Chains sharing the same linkage types can still have different architectures and thus different structures, resulting in a huge number of possible conjugate combinations affecting different signaling pathways (reviewed in [130–132]). Due to their architecture, the study of heterotypic chains represents a challenge, and their functions in cells are still not clear (**Figure 2**).

K11/K48, K48/K63, and M1/K63 are three of the most studied branched chains.

K11 and K48 both target proteins for degradation. Branched K11/K48 ubiquitin chains seem to increase this signal leading to a more efficient recognition of substrates by the proteasome. The APC/C complex assembles K11 chains to substrates, targeting them for protein degradation. During mitosis, the APC/C complex conjugates K11/K48 branched chains to the kinase Nek2A leading to a more efficient recognition by the proteasome for degradation [133]. The binding efficiency of homotypic K11 chains to the proteasome is much lower than that of homotypic K48 chains and heterotypic K48/K11 chains, and that both K48 chains and K11/K48 chains efficiently target cyclin B1 for proteasome degradation [134]. The development of a bispecific antibody to K11/K48 chains allowed the detection of APC/C complex assembling K11/K48 chains during mitosis. Under proteotoxic stress, leading to the accumulation of newly synthesized and misfolded proteins, these linkages seem to accumulate. These chains seem to have a quality control role where they prevent protein aggregation by proteasomal degradation. Among the effectors of these chains are endogenous p97, BAG6, UBQLN2, p62, UBR5, and HUWE1 [135]. More recently, the structure of branched K11/K48-linked tri-ubiquitin was solved, showing the presence of a novel binding surface exclusive to branched K11/K48-linked polyUb as a result of a unique interface between the branched K11 and K48. This interface binds to Rpn1, one of the proteasomal units able to recognize polyUb, and increases binding efficiency [136].

Opposed to K48, K63 chains are non-proteolytic chains playing important roles in different signaling pathways. The combination of K48 and K63 chains in branched chains seems to play an important role in NF-κB signaling [137]. Induced by IL-1βsignaling, HUWE1 cooperates with K63 ubiquitinated TRAF6 to assemble K48 chains to the previously assembled K63 chains. The addition of the K48 chains does not interfere with the recognition by TAB2 but protects K63 chains from deubiquitination by CYLD, enhancing the NF-κB signaling [137]. Interestingly, K63 branched chains seem to target proteins for proteasome degradation. The ubiquitin ligase ITCH is involved in apoptosis regulation through the ubiquitination of TXNIP [138]. ITCH assembles K63 chains to TXNIP that act as a recruitment signal for UBR5 which then assembles K48 branched chains. Because ITCH is a tumor-promoting and anti-apoptotic factor and TXNIP is a tumor suppressor and pro-apoptotic factor, it is possible that during apoptosis, ITCH accelerates TXNIP degradation counteracting its effects [139].

K63/M1 chains, just as M1 and K63 homotypic chains, seem to play a significant role in the innate immune response. Upon activation of MyD88, TNFR1/TRADD, TLR3/TRIF, and NOD1/RIP2 signaling pathways, the formation of K63/M1 branched chains are induced leading to activation of the IKKs [108, 140]. The inflammation-associated protein A20 has both an OTU-type deubiquitinase domain and a ZnF4 motif that binds ubiquitin [141]. After phosphorylation, A20 promotes disassembly of K63 chains during TNFR1 signaling leading to cell death. However, the second step of linear ubiquitination forming branched chains protects TNFR1-associated proteins from K63 disassembling, maintaining the signaling and leading to inflammation [142].

Several other heterotypic chains were found to date. **Table 2** gives an overview of those chains and their biological significance.

3.3 Ubl/ubiquitin chains

Ubiquitin chains can also be modified by ubiquitin-like modifiers (**Figure 2**) [149–153], and although not much is known about these chains, they increase even more the complexity of the ubiquitin code. Ubiquitin chains have been found to be modified by SUMO at K6, K11, K27, K48, and K63. Despite the unclear biological

Chain	Function	Reference
K11/K48	Proteasomal degradation	[133–136]
K48/K63	NF-κB signaling apoptosis	[137, 139]
K63/M1	Innate immune response	[108, 140, 142]
K6/K48	?	[143]
K29/K48	Proteasomal degradation	[55, 144–146]
K11/K63	Endocytosis	[147]
K29/K33	?	[148]

Table 2.
Branched ubiquitin chain types and their associated cellular functions.

role of these modifications, under proteasome inhibition or heat shock conditions, K6 and K27 chains seem to accumulate [151]. ISG15 can form mixed chains with ubiquitin at K29. These chains have a non-proteolytic function and seem to regulate the turnover of ubiquitylated proteins [152]. NEDD8 was found to form branched chains with K48 in human cells acting as a chain terminator [153].

4. Detection methods of ubiquitinated target and chains

4.1 Biochemical and genetic methods

Over the years, different biochemical and genetic methods were developed to detect ubiquitin, ubiquitin-like modifiers, and ubiquitin chains. Although antibodies were available from early on, problems with specificity led to the use of different N-terminally epitope-tagged forms of ubiquitin. Here, antibodies against the epitope tag were used for detection. These constructs were elegantly combined with molecular biological methods, which replaced single lysine residues in ubiquitin with arginine, preventing the formation of ubiquitin chains on these positions. A loss of the chain signal was interpreted as the specific modification by ubiquitin chain with a specific topology. (For a more comprehensive overview see [154]).

For the enrichment of ubiquitin chains of a specific type, biochemical methods and specific antibodies have been developed. While the antibodies were used with varying success due to specificity problems, the use of tandem-repeated ubiquitin-binding entities (TUBE) constructs has gained importance. Here ubiquitin-binding domains with specificity for certain chain topologies are multimerized and used as a pull-down construct [155].

Ubiquitin chain restriction (UbiCRest) is another alternative to identify polyubiquitin chains [76, 156]. In this approach, ovarian tumor family deubiquitinases are incubated with substrates and used as restriction enzymes to detect linkage sites and determine the relative abundance of Ub chains on substrates [76, 156].

4.2 Mass spectrometry-based methods

Mass spectrometry-based proteomics has been used to detect ubiquitination sites for almost 20 years. In 2001, Peng et al. reported that ubiquitinated peptides have a 114 Da mass shift due to the diglycine left behind on a lysine sidechain of ubiquitin from another ubiquitin, after trypsin digestion [157]. In 2003 the same authors, applying their rationale, identified more than 70 ubiquitin-conjugated proteins and 7 ubiquitination sites (K6, K11, K27, K29, K33, K48, and K63) in ubiquitin itself, being 4 of these sites reported for the first time in vivo [158]. In 2009, Tokunaga et al. reported the identification of linear polyubiquitin in NEMO by mass spectrometry, showing that the NF-κB activation pathway is regulated by LUBAC through the polyubiquitination of NEMO [106]. Due to the low abundance of modified peptides in samples, several enrichment strategies were developed to enrich ubiquitinated peptides and improve ubiquitination identification. The development of an antibody against diglycine linked to the ε-amino group of lysine opened the door to the large-scale identification of ubiquitinated substrate proteins [159]. Although the approach was used very successfully by several laboratories leading to the identification of close to 20,000 ubiquitination sites [160–162] it has some drawbacks. One is that the diglycine remnant left by ubiquitin is not unique, and both NEDD8 and ISG15 leave an identical remnant after trypsin digestion. Additionally, the antibody does not recognize linear ubiquitination. Recently, Akimov et al. developed a new antibody specific to a remnant four-mer peptide of the ubiquitin C-terminus after LysC digestion, identifying 60,000 ubiquitination sites [163]. Other relative quantification methods like stable isotope labeling by/with amino acids in cell culture (SILAC), tandem mass tags (TMT), and label-free quantification have been successfully applied [160, 164–166]; however, these methods are using data-dependent measurements, and although they are most suited for PTMs discovery, they cannot provide information on absolute quantities and/or stoichiometry information.

4.3 Ubiquitin topology analysis

While discovery proteomics based on data-dependent methods (DDA) allows the identification of new proteins and can detect the presence of posttranslational modifications, the reliable identification and quantification of peptides in several samples is hampered by the way how peptides are selected by the mass spectrometer for sequencing. Here, the most intense ion at a very moment is selected, which can lead to the selection of different peptides in consecutive mass spectrometric analysis runs. Contrary to discovery proteomics methods, targeted proteomics tries to identify a given set of peptides in every sample, making this method particularly suited for the analysis of ubiquitin topology experiments [167–169].

The most common techniques for targeted proteomics are selected reaction monitoring (SRM) and parallel reaction monitoring (PRM). Both methods have specific requirements for the mass spectrometer used for the analysis. SRM is bound to a triple-quadrupole mass spectrometer, while PRM measurements require an Orbitrap mass spectrometer. For both methods, a list of peptides is preselected, and the corresponding masses are selected continuously. The selected masses are fragmented in a collision cell, and the fragment masses are monitored. In the case of the ubiquitin topology analysis, the key peptides for the ubiquitin chains are targeted [168]. By comparing this signal with its isotope-labeled version spiked into the mix at a known concentration, it is possible to determine the concentration of the peptide [170]. For the analysis of ubiquitin chain topology, the analysis is focused on unique peptides for each of the ubiquitin chain topologies. By digesting a ubiquitin chain with trypsin under denaturing conditions, ubiquitin peptides are generated. Ubiquitin carries an arginine residue at position 74. Trypsin cleaves after this residue and leaves the double glycine residue on the lysine side chain. This creates a branched peptide with an additional mass of 114 Da at the point of modification. For the targeted analysis are seven branched peptides—for each lysine in ubiquitin one—selected and monitored either by SRM or PRM (reviewed in [154, 171]).

4.4 Detection of branched chains

The detection of heterotypic chains represents a challenge. Specific antibodies can only recognize ubiquitin or one chain at a time [172–174]. Mass spectrometry-based proteomics methodologies are based on the digestion of proteins with a protease, generally trypsin, which cuts after lysines or arginines [175]. Branched chains harbor two or more lysines that are ubiquitinated. Detection of double-ubiquitinated ubiquitin is difficult due to two reasons: first, if branch points are separated by several lysines in ubiquitin, the two branch points are separated into two separate peptides, leading to a loss of the information of the double modification. Second, if the two branch points are on adjacent lysines, the resulting peptide is too long to be measured on a mass spectrometer. An alternative is the coupling of antibody-based enrichment with mass spectrometric analysis [80, 155, 159]. Limited proteolysis associated with middle-down proteomics has been used to characterize polyubiquitin chain structures [143, 145]. Top-down proteomics associated with ultraviolet photodissociation was also applied to determine polyubiquitin chain topology [176]. In 2014, Meyer et al. inserted a TEV cleavage site in ubiquitin which after cleavage allowed for the discrimination between branched and unbranched chains [133]. The authors showed that APC/C synthesizes branched chains that enhance proteasome degradation. Later in 2016, Ohtake et al. used a mutagenesis approach to detect K48/K63 branched chains [137]. The authors replaced ubiquitin's arginine 54 by alanine allowing detection of these chains by mass spectrometry and showed how K48 branched chains protect the deubiquitination of K63. In 2017, Yau et al. developed bispecific antibodies against K11/K48 chains, showing their enhanced signal for protein degradation [135]. More recently, Swatek et al. showed that the leader protease of foot and mouth disease virus cleaves di-ubiquitin between arginine 74 and the C-terminal diglycine, originating one truncated ubiquitin (residues 1–74) and a diglycine modified ubiquitin (residues 1–74) [177]. The authors used the approach coupled to intact MS to identify and quantify ubiquitin chains with one, two, or three branches in whole-cell lysates.

5. Conclusion

Over the last decades, the ubiquitin signaling system has been further and further probed, and today it is clear that it is one of the most complex posttranslational cellular signaling systems. It is involved in almost all cellular processes, and the possibilities in terms of signaling are staggering. Understanding the different signals, which are coded in the ubiquitin chains, is one of the biggest challenges of the ubiquitin field, and the identification of branched and mixed chains and the cross talk with the universe of ubiquitin-like modifiers poses even more challenges. The new tools which are becoming available in combination with new mass spectrometric analysis tools will open the horizon for even more layers of signaling and promise to unravel this hidden chapter of biology.

Acknowledgments

GD was supported by FNR Core grant PrismaHIF (C17/BM/11642138).

Conflict of interest

The authors declare no conflict of interest.

Author details

Daniel Perez-Hernandez[1,2], Marta L. Mendes[1] and Gunnar Dittmar[1,3*]

1 Proteomics of Cellular Signaling, Quantitative Biology Unit, Luxembourg Institute of Health, Luxembourg

2 Axel Semrau GmbH & Co. KG, Sprockhövel, Germany

3 Faculty of Science, Technology and Communication, University of Luxembourg, Luxembourg

*Address all correspondence to: gunnar.dittmar@lih.lu

References

[1] Finley D, Bartel B, Varshavsky A. The tails of ubiquitin precursors are ribosomal proteins whose fusion to ubiquitin facilitates ribosome biogenesis. Nature. 1989;**338**:394-401

[2] Varadan R, Walker O, Pickart C, Fushman D. Structural properties of polyubiquitin chains in solution. Journal of Molecular Biology. 2002;**324**: 637-647

[3] Komander D, Reyes-Turcu F, Licchesi JDF, Odenwaelder P, Wilkinson KD, Barford D. Molecular discrimination of structurally equivalent Lys 63-linked and linear polyubiquitin chains. EMBO Reports. 2009;**10**:466-473

[4] Virdee S, Ye Y, Nguyen DP, Komander D, Chin JW. Engineered diubiquitin synthesis reveals Lys29-isopeptide specificity of an OTU deubiquitinase. Nature Chemical Biology. 2010;**6**:750-757

[5] Varadan R, Assfalg M, Haririnia A, Raasi S, Pickart C, Fushman D. Solution conformation of Lys63-linked di-ubiquitin chain provides clues to functional diversity of polyubiquitin signaling. The Journal of Biological Chemistry. 2004;**279**:7055-7063

[6] Tenno T, Fujiwara K, Tochio H, Iwai K, Hayato Morita E, Hayashi H, et al. Structural basis for distinct roles of Lys63- and Lys48-linked polyubiquitin chains. Genes to Cells. 2004:865-875

[7] Mevissen TET, Komander D. Mechanisms of Deubiquitinase specificity and regulation. Annual Review of Biochemistry. 2017;**86**:159-192

[8] Liakopoulos D, Doenges G, Matuschewski K, Jentsch S. A novel protein modification pathway related to the ubiquitin system. The EMBO Journal. 1998;**17**:2208-2214

[9] Enchev RI, Schulman BA, Peter M. Protein neddylation: Beyond cullin-RING ligases. Nature Reviews. Molecular Cell Biology. 2015;**16**:30-44

[10] Scott DC, Sviderskiy VO, Monda JK, Lydeard JR, Cho SE, Harper JW, et al. Structure of a RING E3 trapped in action reveals ligation mechanism for the ubiquitin-like protein NEDD8. Cell. 2014;**157**:1671-1684

[11] Wolf DA, Zhou C, Wee S. The COP9 signalosome: An assembly and maintenance platform for cullin ubiquitin ligases? Nature Cell Biology. 2003;**5**:1029-1033

[12] Cope GA, Suh GSB, Aravind L, Schwarz SE, Zipursky SL, Koonin EV, et al. Role of predicted metalloprotease motif of Jab1/Csn5 in cleavage of Nedd8 from Cul1. Science. 2002;**298**:608-611

[13] Lyapina S, Cope G, Shevchenko A, Serino G, Tsuge T, Zhou C, et al. Promotion of NEDD-CUL1 conjugate cleavage by COP9 signalosome. Science. 2001;**292**:1382-1385

[14] Zhou C, Seibert V, Geyer R, Rhee E, Lyapina S, Cope G, et al. The fission yeast COP9/signalosome is involved in cullin modification by ubiquitin-related Ned8p. BMC Biochemistry. 2001;**2**:7

[15] Loftus SJ, Liu G, Carr SM, Munro S, La Thangue NB. NEDDylation regulates E2F-1-dependent transcription. EMBO Reports. 2012;**13**:811-818

[16] Xirodimas DP, Saville MK, Bourdon J-C, Hay RT, Lane DP. Mdm2-mediated NEDD8 conjugation of p53 inhibits its transcriptional activity. Cell. 2004;**118**:83-97

[17] Dohmesen C, Koeppel M, Dobbelstein M. Specific inhibition of Mdm2-mediated neddylation by Tip60. Cell Cycle. 2008;**7**:222-231

[18] Schwarz SE, Matuschewski K, Liakopoulos D, Scheffner M, Jentsch S. The ubiquitin-like proteins SMT3 and SUMO-1 are conjugated by the UBC9 E2 enzyme. Proceedings of the National Academy of Sciences of the United States of America. 1998;**95**:560-564

[19] Qin Y, Bao H, Pan Y, Yin M, Liu Y, Wu S, et al. SUMOylation alterations are associated with multidrug resistance in hepatocellular carcinoma. Molecular Medicine Reports. 2014;**9**:877-881

[20] Zhu S, Sachdeva M, Wu F, Lu Z, Mo Y. Ubc9 promotes breast cell invasion and metastasis in a sumoylation-independent manner. Oncogene. 2010;**29**:1763-1772

[21] Lin D, Tatham MH, Yu B, Kim S, Hay RT, Chen Y. Identification of a substrate recognition site on Ubc9. The Journal of Biological Chemistry. 2002;**277**:21740-21748

[22] Bernier-Villamor V, Sampson DA, Matunis MJ, Lima CD. Structural basis for E2-mediated SUMO conjugation revealed by a complex between ubiquitin-conjugating enzyme Ubc9 and RanGAP1. Cell. 2002;**108**:345-356

[23] Reverter D, Lima CD. Insights into E3 ligase activity revealed by a SUMO-RanGAP1-Ubc9-Nup358 complex. Nature. 2005;**435**:687-692

[24] Pichler A, Knipscheer P, Saitoh H, Sixma TK, Melchior F. The RanBP2 SUMO E3 ligase is neither HECT- nor RING-type. Nature Structural & Molecular Biology. 2004;**11**:984-991

[25] Pichler A, Gast A, Seeler JS, Dejean A, Melchior F. The nucleoporin RanBP2 has SUMO1 E3 ligase activity. Cell. 2002;**108**:109-120

[26] Pichler A, Fatouros C, Lee H, Eisenhardt N. SUMO conjugation - a mechanistic view. Biomolecular Concepts. 2017;**8**:13-36

[27] Golebiowski F, Matic I, Tatham MH, Cole C, Yin Y, Nakamura A, et al. System-wide changes to SUMO modifications in response to heat shock. Science Signaling. 2009;**2**:ra24

[28] Matunis MJ, Wu J, Blobel G. SUMO-1 modification and its role in targeting the ran GTPase-activating protein, RanGAP1, to the nuclear pore complex. The Journal of Cell Biology. 1998;**140**:499-509

[29] Mahajan R, Gerace L, Melchior F. Molecular characterization of the SUMO-1 modification of RanGAP1 and its role in nuclear envelope association. The Journal of Cell Biology. 1998;**140**:259-270

[30] Cubeñas-Potts C, Matunis MJ. SUMO: A multifaceted modifier of chromatin structure and function. Developmental Cell. 2013;**24**:1-12

[31] Nie M, Boddy MN. Cooperativity of the SUMO and ubiquitin pathways in genome stability. Biomolecules. 2016;**6**:14

[32] Flotho A, Melchior F. Sumoylation: A regulatory protein modification in health and disease. Annual Review of Biochemistry. 2013;**8**:357-385

[33] Hendriks IA, Vertegaal ACO. A comprehensive compilation of SUMO proteomics. Nature Reviews. Molecular Cell Biology. 2016;**17**:581-595

[34] Dzimianski JV, Scholte FEM, Bergeron É, Pegan SD. ISG15: It's complicated. Journal of Molecular Biology. 2019:4203-4216

[35] Dastur A, Beaudenon S, Kelley M, Krug RM, Huibregtse JM. Herc5, an interferon-induced HECT E3 enzyme, is required for conjugation of ISG15 in human cells. The Journal of Biological Chemistry. 2006;**281**:4334-4338

[36] Malakhov MP, Malakhova OA, Kim KI, Ritchie KJ, Zhang D-E. UBP43

(USP18) specifically removes ISG15 from conjugated proteins. The Journal of Biological Chemistry. 2002;**277**:9976-9981

[37] Nakashima H, Nguyen T, Goins WF, Chiocca EA. Interferon-stimulated gene 15 (ISG15) and ISG15-linked proteins can associate with members of the selective autophagic process, histone deacetylase 6 (HDAC6) and SQSTM1/p62. The Journal of Biological Chemistry. 2015;**290**:1485-1495

[38] Villarroya-Beltri C, Baixauli F, Mittelbrunn M, Fernández-Delgado I, Torralba D, Moreno-Gonzalo O, et al. ISGylation controls exosome secretion by promoting lysosomal degradation of MVB proteins. Nature Communications. 2016;**7**:13588

[39] Yeh Y-H, Yang Y-C, Hsieh M-Y, Yeh Y-C, Li T-K. A negative feedback of the HIF-1α pathway via interferon-stimulated gene 15 and ISGylation. Clinical Cancer Research. 2013;**19**:5927-5939

[40] Fan J-B, Miyauchi-Ishida S, Arimoto K-I, Liu D, Yan M, Liu C-W, et al. Type I IFN induces protein ISGylation to enhance cytokine expression and augments colonic inflammation. Proceedings of the National Academy of Sciences of the United States of America. 2015;**112**:14313-14318

[41] Park JM, Yang SW, Yu KR, Ka SH, Lee SW, Seol JH, et al. Modification of PCNA by ISG15 plays a crucial role in termination of error-prone translesion DNA synthesis. Molecular Cell. 2014:626-638

[42] Okumura F, Okumura AJ, Uematsu K, Hatakeyama S, Zhang D-E, Kamura T. Activation of double-stranded RNA-activated protein kinase (PKR) by interferon-stimulated gene 15 (ISG15) modification down-regulates protein translation. Journal of Biological Chemistry. 2013;**288**:2839-2847

[43] Zhao C, Denison C, Huibregtse JM, Gygi S, Krug RM. Human ISG15 conjugation targets both IFN-induced and constitutively expressed proteins functioning in diverse cellular pathways. Proceedings of the National Academy of Sciences of the United States of America. 2005;**102**:10200-10205

[44] Hsiang T-Y, Zhao C, Krug RM. Interferon-induced ISG15 conjugation inhibits influenza a virus gene expression and replication in human cells. Journal of Virology. 2009;**83**:5971-5977

[45] Lenschow DJ, Giannakopoulos NV, Gunn LJ, Johnston C, O'Guin AK, Schmidt RE, et al. Identification of interferon-stimulated gene 15 as an antiviral molecule during Sindbis virus infection in vivo. Journal of Virology. 2005;**79**:13974-13983

[46] Okumura A, Lu G, Pitha-Rowe I, Pitha PM. Innate antiviral response targets HIV-1 release by the induction of ubiquitin-like protein ISG15. Proceedings of the National Academy of Sciences of the United States of America. 2006;**103**:1440-1445

[47] Okumura A, Pitha PM, Harty RN. ISG15 inhibits Ebola VP40 VLP budding in an L-domain-dependent manner by blocking Nedd4 ligase activity. Proceedings of the National Academy of Sciences. 2008;**105**:3974-3979

[48] Shi H-X, Yang K, Liu X, Liu X-Y, Wei B, Shan Y-F, et al. Positive regulation of interferon regulatory factor 3 activation by Herc5 via ISG15 modification. Molecular and Cellular Biology. 2010;**30**:2424-2436

[49] Kim YJ, Kim ET, Kim Y-E, Lee MK, Kwon KM, Kim KI, et al. Consecutive inhibition of ISG15 expression and ISGylation by cytomegalovirus regulators. PLoS Pathogens. 2016;**12**:e1005850

[50] González-Sanz R, Mata M, Bermejo-Martín J, Álvarez A, Cortijo J, Melero JA, et al. ISG15 is upregulated in respiratory syncytial virus infection and reduces virus growth through protein ISGylation. Journal of Virology. 2016;**90**:3428-3438

[51] Terrell J, Shih S, Dunn R, Hicke L. A function for monoubiquitination in the internalization of a G protein-coupled receptor. Molecular Cell. 1998;**1**:193-202

[52] Haglund K, Sigismund S, Polo S, Szymkiewicz I, Di Fiore PP, Dikic I. Multiple monoubiquitination of RTKs is sufficient for their endocytosis and degradation. Nature Cell Biology. 2003;**5**:461-466

[53] Ye Y, Blaser G, Horrocks MH, Ruedas-Rama MJ, Ibrahim S, Zhukov AA, et al. Ubiquitin chain conformation regulates recognition and activity of interacting proteins. Nature. 2012;**492**:266-270

[54] Bremm A, Freund SMV, Komander D. Lys11-linked ubiquitin chains adopt compact conformations and are preferentially hydrolyzed by the deubiquitinase Cezanne. Nature Structural & Molecular Biology. 2010;**17**:939-947

[55] Kristariyanto YA, Abdul Rehman SA, Campbell DG, Morrice NA, Johnson C, Toth R, et al. K29-selective ubiquitin binding domain reveals structural basis of specificity and heterotypic nature of k29 polyubiquitin. Molecular Cell. 2015;**58**:83-94

[56] Castañeda CA, Dixon EK, Walker O, Chaturvedi A, Nakasone MA, Curtis JE, et al. Linkage via K27 bestows ubiquitin chains with unique properties among Polyubiquitins. Structure. 2016;**24**:423-436

[57] Michel MA, Elliott PR, Swatek KN, Simicek M, Pruneda JN, Wagstaff JL, et al. Assembly and specific recognition of k29- and k33-linked polyubiquitin. Molecular Cell. 2015;**58**:95-109

[58] Liu Z, Gong Z, Jiang W-X, Yang J, Zhu W-K, Guo D-C, et al. Lys63-linked ubiquitin chain adopts multiple conformational states for specific target recognition. eLife. 2015;**4**:1-19

[59] Zhang X, Smits AH, van Tilburg GBA, Jansen PWTC, Makowski MM, Ovaa H, et al. An interaction landscape of ubiquitin signaling. Molecular Cell. 2017;**65**: 941-955.e8

[60] Dammer EB, Na CH, Xu P, Seyfried NT, Duong DM, Cheng D, et al. Polyubiquitin linkage profiles in three models of proteolytic stress suggest the etiology of Alzheimer disease. The Journal of Biological Chemistry. 2011;**286**:10457-10465

[61] Ziv I, Matiuhin Y, Kirkpatrick DS, Erpapazoglou Z, Leon S, Pantazopoulou M, et al. A perturbed ubiquitin landscape distinguishes between ubiquitin in trafficking and in proteolysis. Molecular & Cellular Proteomics. 2011;**10**:M111.009753

[62] Cunningham CN, Baughman JM, Phu L, Tea JS, Yu C, Coons M, et al. USP30 and parkin homeostatically regulate atypical ubiquitin chains on mitochondria. Nature Cell Biology. 2015;**17**:160-169

[63] Ordureau A, Sarraf SA, Duda DM, Heo J-M, Jedrychowski MP, Sviderskiy VO, et al. Quantitative proteomics reveal a feedforward mechanism for mitochondrial PARKIN translocation and ubiquitin chain synthesis. Molecular Cell. 2014;**56**:360-375

[64] Gersch M, Gladkova C, Schubert AF, Michel MA, Maslen S, Komander D. Mechanism and regulation of the Lys6-selective deubiquitinase USP30. Nature Structural & Molecular Biology. 2017;**24**:920-930

[65] Elia AEH, Boardman AP, Wang DC, Huttlin EL, Everley RA, Dephoure N, et al. Quantitative proteomic atlas of Ubiquitination and acetylation in the DNA damage response. Molecular Cell. 2015;**59**:867-881

[66] Kao S-H, Wu H-T, Wu K-J. Ubiquitination by HUWE1 in tumorigenesis and beyond. Journal of Biomedical Science. 2018;**25**:67

[67] Strappazzon F, Di Rita A, Peschiaroli A, Leoncini PP, Locatelli F, Melino G, et al. HUWE1 controls MCL1 stability to unleash AMBRA1-induced mitophagy. Cell Death and Differentiation. 2019. Epub ahead of print

[68] Michel MA, Swatek KN, Hospenthal MK, Komander D. Ubiquitin linkage-specific affimers reveal insights into K6-linked ubiquitin signaling. Molecular Cell. 2017;**68**:233-246.e5

[69] Palikaras K, Lionaki E, Tavernarakis N. Mechanisms of mitophagy in cellular homeostasis, physiology and pathology. Nature Cell Biology. 2018;**20**:1013-1022

[70] Kitada T, Asakawa S, Hattori N, Matsumine H, Yamamura Y, Minoshima S, et al. Mutations in the parkin gene cause autosomal recessive juvenile parkinsonism. Nature. 1998;**392**:605-608

[71] Valente EM, Abou-Sleiman PM, Caputo V, Muqit MMK, Harvey K, Gispert S, et al. Hereditary early-onset Parkinson's disease caused by mutations in PINK1. Science. 2004;**304**:1158-1160

[72] Kirkpatrick DS, Hathaway NA, Hanna J, Elsasser S, Rush J, Finley D, et al. Quantitative analysis of in vitro ubiquitinated cyclin B1 reveals complex chain topology. Nature Cell Biology. 2006;**8**:700-710

[73] Jin L, Williamson A, Banerjee S, Philipp I, Rape M. Mechanism of ubiquitin-chain formation by the human anaphase-promoting complex. Cell. 2008;**133**:653-665

[74] Matsumoto ML, Wickliffe KE, Dong KC, Yu C, Bosanac I, Bustos D, et al. K11-linked polyubiquitination in cell cycle control revealed by a K11 linkage-specific antibody. Molecular Cell. 2010;**39**:477-484

[75] Min M, Mevissen TET, De Luca M, Komander D, Lindon C. Efficient APC/C substrate degradation in cells undergoing mitotic exit depends on K11 ubiquitin linkages. Molecular Biology of the Cell. 2015;**26**:4325-4332

[76] Mevissen TET, Hospenthal MK, Geurink PP, Elliott PR, Akutsu M, Arnaudo N, et al. OTU deubiquitinases reveal mechanisms of linkage specificity and enable ubiquitin chain restriction analysis. Cell. 2013;**154**:169-184

[77] Bremm A, Moniz S, Mader J, Rocha S, Komander D. Cezanne (OTUD7B) regulates HIF-1α homeostasis in a proteasome-independent manner. EMBO Reports. 2014;**15**:1268-1277

[78] Bonacci T, Suzuki A, Grant GD, Stanley N, Cook JG, Brown NG, et al. Cezanne/OTUD 7B is a cell cycle-regulated deubiquitinase that antagonizes the degradation of APC/C substrates. The EMBO Journal. 2018;**37**(16):1-17

[79] Li Y, Dammer EB, Gao Y, Lan Q, Villamil MA, Duong DM, et al. Proteomics links ubiquitin chain topology change to transcription factor activation. Molecular Cell. 2019;**76**: 126-137.e7

[80] Flick K, Ouni I, Wohlschlegel JA, Capati C, McDonald WH, Yates JR, et al. Proteolysis-independent regulation of the transcription factor Met4 by a single Lys 48-linked ubiquitin chain. Nature Cell Biology. 2004;**6**:634-641

[81] Liu Z, Chen P, Gao H, Gu Y, Yang J, Peng H, et al. Ubiquitylation of autophagy receptor Optineurin by HACE1 activates selective autophagy for tumor suppression. Cancer Cell. 2014;**26**:106-120

[82] Palicharla VR, Maddika S. HACE1 mediated K27 ubiquitin linkage leads to YB-1 protein secretion. Cellular Signalling. 2015;**27**:2355-2362

[83] Gatti M, Pinato S, Maiolica A, Rocchio F, Prato MG, Aebersold R, et al. RNF168 promotes noncanonical K27 ubiquitination to signal DNA damage. Cell Reports. 2015;**10**:226-238

[84] Wang Q, Liu X, Cui Y, Tang Y, Chen W, Li S, et al. The E3 ubiquitin ligase AMFR and INSIG1 bridge the activation of TBK1 kinase by modifying the adaptor STING. Immunity. 2014;**41**:919-933

[85] Xue B, Li H, Guo M, Wang J, Xu Y, Zou X, et al. TRIM21 promotes innate immune response to RNA viral infection through Lys27-linked polyubiquitination of MAVS. Journal of Virology. 2018;**92**(14):1-19

[86] Aki D, Li H, Zhang W, Zheng M, Elly C, Lee JH, et al. The E3 ligases itch and WWP2 cooperate to limit TH2 differentiation by enhancing signaling through the TCR. Nature Immunology. 2018;**19**:766-775

[87] Wu X, Lei C, Xia T, Zhong X, Yang Q, Shu H-B. Regulation of TRIF-mediated innate immune response by K27-linked polyubiquitination and deubiquitination. Nature Communications. 2019;**10**:4115

[88] Cho JJ, Xu Z, Parthasarathy U, Drashansky TT, Helm EY, Zuniga AN, et al. Hectd3 promotes pathogenic Th17 lineage through Stat3 activation and Malt1 signaling in neuroinflammation. Nature Communications. 2019;**10**:701

[89] Hay-Koren A, Caspi M, Zilberberg A, Rosin-Arbesfeld R. The EDD E3 ubiquitin ligase ubiquitinates and up-regulates β-catenin. Molecular Biology of the Cell (MBoC). 2011;**22**:399-411

[90] Fei C, Li Z, Li C, Chen Y, Chen Z, He X, et al. Smurf1-mediated Lys29-linked nonproteolytic polyubiquitination of axin negatively regulates Wnt/β-catenin signaling. Molecular and Cellular Biology. 2013;**33**:4095-4105

[91] Licchesi JDF, Mieszczanek J, Mevissen TET, Rutherford TJ, Akutsu M, Virdee S, et al. An ankyrin-repeat ubiquitin-binding domain determines TRABID's specificity for atypical ubiquitin chains. Nature Structural & Molecular Biology. 2011;**19**:62-71

[92] Tran H, Hamada F, Schwarz-Romond T, Bienz M. Trabid, a new positive regulator of Wnt-induced transcription with preference for binding and cleaving K63-linked ubiquitin chains. Genes & Development. 2008;**22**:528-542

[93] Jin J, Xie X, Xiao Y, Hu H, Zou Q, Cheng X, et al. Epigenetic regulation of the expression of Il12 and Il23 and autoimmune inflammation by the deubiquitinase Trabid. Nature Immunology. 2016;**17**:259-268

[94] Huang H, Jeon M-S, Liao L, Yang C, Elly C, Yates JR 3rd, et al. K33-linked polyubiquitination of T cell receptor-zeta regulates proteolysis-independent T cell signaling. Immunity. 2010;**33**:60-70

[95] Lin M, Zhao Z, Yang Z, Meng Q, Tan P, Xie W, et al. USP38 inhibits type I interferon signaling by editing TBK1 Ubiquitination through NLRP4 signalosome. Molecular Cell. 2016;**64**:267-281

[96] Miao Y, Wu J, Abraham SN. Ubiquitination of innate immune regulator TRAF3 orchestrates expulsion

of intracellular Bacteria by exocyst complex. Immunity. 2016;**45**:94-105

[97] Zhao Y, Wang X, Wang Q, Deng Y, Li K, Zhang M, et al. USP2a supports metastasis by tuning TGF-β signaling. Cell Reports. 2018;**22**:2442-2454

[98] Yuan W-C, Lee Y-R, Lin S-Y, Chang L-Y, Tan YP, Hung C-C, et al. K33-linked polyubiquitination of Coronin 7 by Cul3-KLHL20 ubiquitin E3 ligase regulates protein trafficking. Molecular Cell. 2014;**54**:586-600

[99] Heath RJ, Goel G, Baxt LA, Rush JS, Mohanan V, Paulus GLC, et al. RNF166 determines recruitment of adaptor proteins during antibacterial autophagy. Cell Reports. 2016;**17**:2183-2194

[100] Feng X, Jia Y, Zhang Y, Ma F, Zhu Y, Hong X, et al. Ubiquitination of UVRAG by SMURF1 promotes autophagosome maturation and inhibits hepatocellular carcinoma growth. Autophagy. 2019;**15**:1130-1149

[101] Walczak H, Iwai K, Dikic I. Generation and physiological roles of linear ubiquitin chains. BMC Biology. 2012;**10**:23

[102] Dittmar G, Winklhofer KF. Linear ubiquitin chains: Cellular functions and strategies for detection and quantification. Frontiers in Chemistry. 2020;**7**:915

[103] Kirisako T, Kamei K, Murata S, Kato M, Fukumoto H, Kanie M, et al. A ubiquitin ligase complex assembles linear polyubiquitin chains. The EMBO Journal. 2006;**25**:4877-4887

[104] Ikeda F, Deribe YL, Skånland SS, Stieglitz B, Grabbe C, Franz-Wachtel M, et al. SHARPIN forms a linear ubiquitin ligase complex regulating NF-κB activity and apoptosis. Nature. 2011;**471**:637-641

[105] Gerlach B, Cordier SM, Schmukle AC, Emmerich CH, Rieser E, Haas TL, et al. Linear ubiquitination prevents inflammation and regulates immune signalling. Nature. 2011;**471**:591-596

[106] Tokunaga F, Sakata S-I, Saeki Y, Satomi Y, Kirisako T, Kamei K, et al. Involvement of linear polyubiquitylation of NEMO in NF-kappaB activation. Nature Cell Biology. 2009;**11**:123-132

[107] Haas TL, Emmerich CH, Gerlach B, Schmukle AC, Cordier SM, Rieser E, et al. Recruitment of the linear ubiquitin chain assembly complex stabilizes the TNF-R1 signaling complex and is required for TNF-mediated gene induction. Molecular Cell. 2009;**36**:831-844

[108] Emmerich CH, Ordureau A, Strickson S, Arthur JSC, Pedrioli PGA, Komander D, et al. Activation of the canonical IKK complex by K63/M1-linked hybrid ubiquitin chains. Proceedings of the National Academy of Sciences of the United States of America. 2013;**110**:15247-15252

[109] Müller-Rischart AK, Pilsl A, Beaudette P, Patra M, Hadian K, Funke M, et al. The E3 ligase parkin maintains mitochondrial integrity by increasing linear ubiquitination of NEMO. Molecular Cell. 2013;**49**:908-921

[110] Kwon YT, Ciechanover A. The ubiquitin code in the ubiquitin-proteasome system and autophagy. Trends in Biochemical Sciences. 2017;**42**:873-886

[111] Chau V, Tobias JW, Bachmair A, Marriott D, Ecker DJ, Gonda DK, et al. A multiubiquitin chain is confined to specific lysine in a targeted short-lived protein. Science. 1989;**243**:1576-1583

[112] Finley D, Sadis S, Monia BP, Boucher P, Ecker DJ, Crooke ST, et al. Inhibition of proteolysis and cell cycle progression in a multiubiquitination-deficient yeast mutant. Molecular and Cellular Biology. 1994;**14**:5501-5509

[113] Thrower JS, Hoffman L, Rechsteiner M, Pickart CM. Recognition of the polyubiquitin proteolytic signal. The EMBO Journal. 2000;**19**:94-102

[114] Boutet SC, Disatnik M-H, Chan LS, Iori K, Rando TA. Regulation of Pax3 by proteasomal degradation of monoubiquitinated protein in skeletal muscle progenitors. Cell. 2007;**130**:349-362

[115] Kravtsova-Ivantsiv Y, Cohen S, Ciechanover A. Modification by single ubiquitin moieties rather than polyubiquitination is sufficient for proteasomal processing of the p105 NF-kappaB precursor. Molecular Cell. 2009;**33**:496-504

[116] Braten O, Livneh I, Ziv T, Admon A, Kehat I, Caspi LH, et al. Numerous proteins with unique characteristics are degraded by the 26S proteasome following monoubiquitination. Proceedings of the National Academy of Sciences of the United States of America. 2016;**113**:E4639-E4647

[117] Komander D, Rape M. The ubiquitin code. Annual Review of Biochemistry. 2012;**81**:203-229

[118] Swatek KN, Komander D. Ubiquitin modifications. Cell Research. 2016;**26**:399-422

[119] Erpapazoglou Z, Walker O, Haguenauer-Tsapis R. Versatile roles of K63-linked ubiquitin chains in trafficking. Cell. 2014;**3**:1027

[120] Deng L, Wang C, Spencer E, Yang L, Braun A, You J, et al. Activation of the IkappaB kinase complex by TRAF6 requires a dimeric ubiquitin-conjugating enzyme complex and a unique polyubiquitin chain. Cell. 2000;**103**:351-361

[121] Yang W-L, Wang J, Chan C-H, Lee S-W, Campos AD, Lamothe B, et al. The E3 ligase TRAF6 regulates Akt ubiquitination and activation. Science. 2009;**325**:1134-1138

[122] Gack MU, Shin YC, Joo C-H, Urano T, Liang C, Sun L, et al. TRIM25 RING-finger E3 ubiquitin ligase is essential for RIG-I-mediated antiviral activity. Nature. 2007;**446**:916-920

[123] Ordureau A, Smith H, Windheim M, Peggie M, Carrick E, Morrice N, et al. The IRAK-catalysed activation of the E3 ligase function of Pellino isoforms induces the Lys63-linked polyubiquitination of IRAK1. The Biochemical Journal. 2008;**409**:43-52

[124] Tsuchida T, Zou J, Saitoh T, Kumar H, Abe T, Matsuura Y, et al. The ubiquitin ligase TRIM56 regulates innate immune responses to intracellular double-stranded DNA. Immunity. 2010;**33**:765-776

[125] Schwertman P, Bekker-Jensen S, Mailand N. Regulation of DNA double-strand break repair by ubiquitin and ubiquitin-like modifiers. Nature Reviews. Molecular Cell Biology. 2016;**17**:379-394

[126] Duncan LM, Piper S, Dodd RB, Saville MK, Sanderson CM, Luzio JP, et al. Lysine-63-linked ubiquitination is required for endolysosomal degradation of class I molecules. The EMBO Journal. 2006;**25**:1635-1645

[127] Huang F, Kirkpatrick D, Jiang X, Gygi S, Sorkin A. Differential regulation of EGF receptor internalization and degradation by multiubiquitination within the kinase domain. Molecular Cell. 2006;**21**:737-748

[128] Bertelsen V, Sak MM, Breen K, Rødland MS, Johannessen LE, Traub LM, et al. A chimeric pre-ubiquitinated EGF receptor is constitutively endocytosed in a clathrin-dependent, but kinase-independent manner. Traffic. 2011;**12**:507-520

[129] Liu P, Gan W, Su S, Hauenstein AV, Fu T-M, Brasher B, et al. K63-linked polyubiquitin chains bind to DNA to facilitate DNA damage repair. Science Signaling. 2018;**11**(533):1-12

[130] Yau R, Rape M. The increasing complexity of the ubiquitin code. Nature Cell Biology. 2016;**18**:579-586

[131] Ohtake F, Tsuchiya H. The emerging complexity of ubiquitin architecture. Journal of Biochemistry. 2017;**161**:125-133

[132] Haakonsen DL, Rape M. Branching out: Improved signaling by heterotypic ubiquitin chains. Trends in Cell Biology. 2019;**29**(9):704-716

[133] Meyer H-J, Rape M. Enhanced protein degradation by branched ubiquitin chains. Cell. 2014;**157**:910-921

[134] Grice GL, Lobb IT, Weekes MP, Gygi SP, Antrobus R, Nathan JA. The proteasome distinguishes between heterotypic and Homotypic Lysine-11-linked Polyubiquitin chains. Cell Reports. 2015;**12**:545-553

[135] Yau RG, Doerner K, Castellanos ER, Haakonsen DL, Werner A, Wang N, et al. Assembly and function of heterotypic ubiquitin chains in cell-cycle and protein quality control. Cell. 2017;**171**:918-933.e20

[136] Boughton AJ, Krueger S, Fushman D. Branching via K11 and K48 bestows ubiquitin chains with a unique interdomain interface and enhanced affinity for proteasomal subunit Rpn1. Structure. 2020;**28**:29-43.e6

[137] Ohtake F, Saeki Y, Ishido S, Kanno J, Tanaka K. The K48-K63 branched ubiquitin chain regulates NF-κB signaling. Molecular Cell. 2016;**64**:251-266

[138] Zhang P, Wang C, Gao K, Wang D, Mao J, An J, et al. The ubiquitin ligase itch regulates apoptosis by targeting thioredoxin-interacting protein for ubiquitin-dependent degradation. The Journal of Biological Chemistry. 2010;**285**:8869-8879

[139] Ohtake F, Tsuchiya H, Saeki Y, Tanaka K. K63 ubiquitylation triggers proteasomal degradation by seeding branched ubiquitin chains. Proceedings of the National Academy of Sciences of the United States of America. 2018;**115**:E1401-E1408

[140] Emmerich CH, Bakshi S, Kelsall IR, Ortiz-Guerrero J, Shpiro N, Cohen P. Lys63/Met1-hybrid ubiquitin chains are commonly formed during the activation of innate immune signalling. Biochemical and Biophysical Research Communications. 2016;**474**:452-461

[141] Wertz IE, O'Rourke KM, Zhou H, Eby M, Aravind L, Seshagiri S, et al. De-ubiquitination and ubiquitin ligase domains of A20 downregulate NF-kappaB signalling. Nature. 2004;**430**:694-699

[142] Wertz IE, Newton K, Seshasayee D, Kusam S, Lam C, Zhang J, et al. Phosphorylation and linear ubiquitin direct A20 inhibition of inflammation. Nature. 2015;**528**:370-375

[143] Valkevich EM, Sanchez NA, Ge Y, Strieter ER. Middle-down mass spectrometry enables characterization of branched ubiquitin chains. Biochemistry. 2014;**53**:4979-4989

[144] Liu C, Liu W, Ye Y, Li W. Ufd2p synthesizes branched ubiquitin chains to promote the degradation of substrates modified with atypical chains. Nature Communications. 2017;**8**:14274

[145] Xu P, Peng J. Characterization of polyubiquitin chain structure by middle-down mass spectrometry. Analytical Chemistry. 2008;**80**:3438-3444

[146] Saeki Y, Tayama Y, Toh-e A, Yokosawa H. Definitive evidence for Ufd2-catalyzed elongation of the ubiquitin chain through Lys48 linkage. Biochemical and Biophysical Research Communications. 2004;**320**:840-845

[147] Boname JM, Thomas M, Stagg HR, Xu P, Peng J, Lehner PJ. Efficient internalization of MHC I requires lysine-11 and lysine-63 mixed linkage polyubiquitin chains. Traffic. 2010;**11**:210-220

[148] Hospenthal MK, Freund SMV, Komander D. Assembly, analysis and architecture of atypical ubiquitin chains. Nature Structural & Molecular Biology. 2013;**20**:555-565

[149] Galisson F, Mahrouche L, Courcelles M, Bonneil E, Meloche S, Chelbi-Alix MK, et al. A novel proteomics approach to identify SUMOylated proteins and their modification sites in human cells. Molecular & Cellular Proteomics. 2011;**10**:M110.004796

[150] Lamoliatte F, Bonneil E, Durette C, Caron-Lizotte O, Wildemann D, Zerweck J, et al. Targeted identification of SUMOylation sites in human proteins using affinity enrichment and paralog-specific reporter ions. Molecular & Cellular Proteomics. 2013;**12**:2536-2550

[151] Hendriks IA, D'Souza RCJ, Yang B, Verlaan-de Vries M, Mann M, Vertegaal ACO. Uncovering global SUMOylation signaling networks in a site-specific manner. Nature Structural & Molecular Biology. 2014;**21**:927-936

[152] Fan J-B, Arimoto K-I, Motamedchaboki K, Yan M, Wolf DA, Zhang D-E. Identification and characterization of a novel ISG15-ubiquitin mixed chain and its role in regulating protein homeostasis. Scientific Reports. 2015;**5**:12704

[153] Singh RK, Zerath S, Kleifeld O, Scheffner M, Glickman MH, Fushman D. Recognition and cleavage of related to ubiquitin 1 (Rub1) and Rub1-ubiquitin chains by components of the ubiquitin-proteasome system. Molecular & Cellular Proteomics. 2012;**11**:1595-1611

[154] Beaudette P, Popp O, Dittmar G. Proteomic techniques to probe the ubiquitin landscape. Proteomics. 2016;**16**:273-287

[155] Hjerpe R, Aillet F, Lopitz-Otsoa F, Lang V, England P, Rodriguez MS. Efficient protection and isolation of ubiquitylated proteins using tandem ubiquitin-binding entities. EMBO Reports. 2009;**10**:1250-1258

[156] Hospenthal MK, Mevissen TET, Komander D. Deubiquitinase-based analysis of ubiquitin chain architecture using ubiquitin chain restriction (UbiCRest). Nature Protocols. 2015;**10**:349-361

[157] Peng J, Gygi SP. Proteomics: The move to mixtures. Journal of Mass Spectrometry. 2001;**36**:1083-1091

[158] Peng J, Schwartz D, Elias JE, Thoreen CC, Cheng D, Marsischky G, et al. A proteomics approach to understanding protein ubiquitination. Nature Biotechnology. 2003;**21**:921-926

[159] Xu G, Paige JS, Jaffrey SR. Global analysis of lysine ubiquitination by ubiquitin remnant immunoaffinity profiling. Nature Biotechnology. 2010;**28**:868-873

[160] Wagner SA, Beli P, Weinert BT, Nielsen ML, Cox J, Mann M, et al. A proteome-wide, quantitative survey of in vivo ubiquitylation sites reveals widespread regulatory roles. Molecular & Cellular Proteomics. 2011;**10**:M111.013284

[161] Kim W, Bennett EJ, Huttlin EL, Guo A, Li J, Possemato A, et al. Systematic and quantitative assessment of

the ubiquitin-modified proteome. Molecular Cell. 2011;**44**:325-340

[162] Emanuele MJ, Elia AEH, Xu Q, Thoma CR, Izhar L, Leng Y, et al. Global identification of modular cullin-RING ligase substrates. Cell. 2011;**147**:459-474

[163] Akimov V, Barrio-Hernandez I, Hansen SVF, Hallenborg P, Pedersen A-K, Bekker-Jensen DB, et al. UbiSite approach for comprehensive mapping of lysine and N-terminal ubiquitination sites. Nature Structural & Molecular Biology. 2018;**25**:631-640

[164] Isasa M, Rose CM, Elsasser S, Navarrete-Perea J, Paulo JA, Finley DJ, et al. Multiplexed, proteome-wide protein expression profiling: Yeast Deubiquitylating enzyme knockout strains. Journal of Proteome Research. 2015;**14**:5306-5317

[165] Swaney DL, Beltrao P, Starita L, Guo A, Rush J, Fields S, et al. Global analysis of phosphorylation and ubiquitylation cross-talk in protein degradation. Nature Methods. 2013;**10**:676-682

[166] Ordureau A, Münch C, Harper JW. Quantifying ubiquitin signaling. Molecular Cell. 2015;**58**:660-676

[167] Shi T, Su D, Liu T, Tang K, Camp DG 2nd, Qian W-J, et al. Advancing the sensitivity of selected reaction monitoring-based targeted quantitative proteomics. Proteomics. 2012;**12**:1074-1092

[168] Picotti P, Aebersold R. Selected reaction monitoring-based proteomics: Workflows, potential, pitfalls and future directions. Nature Methods. 2012;**9**:555-566

[169] Gillette MA, Carr SA. Quantitative analysis of peptides and proteins in biomedicine by targeted mass spectrometry. Nature Methods. 2013;**10**:28-34

[170] Picotti P, Lam H, Campbell D, Deutsch EW, Mirzaei H, Ranish J, et al. A database of mass spectrometric assays for the yeast proteome. Nature Methods. 2008;**5**:913-914

[171] Longworth J, Dittmar G. Assessment of ubiquitin chain topology by targeted mass spectrometry. Methods in Molecular Biology. 1977;**2019**:25-34

[172] Fujimuro M, Sawada H, Yokosawa H. Production and characterization of monoclonal antibodies specific to multi-ubiquitin chains of polyubiquitinated proteins. FEBS Letters. 1994;**349**:173-180

[173] Newton K, Matsumoto ML, Wertz IE, Kirkpatrick DS, Lill JR, Tan J, et al. Ubiquitin chain editing revealed by polyubiquitin linkage-specific antibodies. Cell. 2008;**134**:668-678

[174] Hershko A, Heller H, Elias S, Ciechanover A. Components of ubiquitin-protein ligase system. Resolution, affinity purification, and role in protein breakdown. The Journal of Biological Chemistry. 1983;**258**:8206-8214

[175] Shevchenko A, Tomas H, Havlis J, Olsen JV, Mann M. In-gel digestion for mass spectrometric characterization of proteins and proteomes. Nature Protocols. 2006;**1**:2856-2860

[176] Cannon JR, Martinez-Fonts K, Robotham SA, Matouschek A, Brodbelt JS. Top-down 193-nm ultraviolet photodissociation mass spectrometry for simultaneous determination of polyubiquitin chain length and topology. Analytical Chemistry. 2015;**87**:1812-1820

[177] Swatek KN, Usher JL, Kueck AF, Gladkova C, Mevissen TET, Pruneda JN, et al. Insights into ubiquitin chain architecture using Ub-clipping. Nature. 2019;**572**(7770):533-537

Chapter 2

Ubiquitin-Independent Proteasomal Degradation Mediated by Antizyme

Noriyuki Murai

Abstract

Most of the proteins in eukaryotic cells are degraded by the proteasome in an ubiquitin-dependent manner. However, ubiquitin-independent protein degradation pathway by the 26S proteasome exists in the cells. Ornithine decarboxylase (ODC) is a well-known protein that is degraded by the 26S proteasome without ubiquitination. Degradation of ODC requires the protein, "antizyme (AZ)," that is induced by polyamine and binds to the ODC monomer to inhibit ODC activity and target it to the 26S proteasome for proteolytic degradation. Namely, AZ contributes the feedback regulation of intracellular polyamine level. ODC has been considered to be the only protein that AZ binds and accelerates its degradation. However, recently AZ-mediated proteasomal protein degradation will gradually increase. Most recently, we found that one of the antizyme families, AZ2, accelerates c-Myc degradation by the proteasome without ubiquitination. In this chapter, we introduce latest several ubiquitin-independent proteasomal degradation mediated by antizyme.

Keywords: antizyme, ubiquitin-independent degradation, ornithine decarboxylase, 26S proteasome, polyamines, c-Myc

1. Introduction

In eukaryotic cells, intracellular protein degradation is mainly regulated by the ubiquitin-proteasome system, where abnormal and unwanted proteins are targeted by polyubiquitin, which is produced from monoubiquitin by ubiquitin-activating enzyme (E1) and ubiquitin-conjugating enzyme (E2) [1]. The proteins that conjugated polyubiquitin by ubiquitin ligase (E3) are finally targeted to the 26S proteasome [2]. However, there is accumulating evidence that ubiquitin-independent proteasomal protein degradation pathway exists in the cells [3, 4]. Although ubiquitin-dependent proteasomal protein degradation is carried out normally by 26S proteasome, there are many reports that ubiquitin-independent proteasomal protein degradations are executed by the only 20S proteasome without the energy of ATP hydrolysis [4]. Among others, some ubiquitin-independent degradation pathways are known to be carried out using not the 20S but the 26S proteasome with the energy of ATP hydrolysis. In this chapter, we introduce ubiquitin-independent proteasomal degradation pathway mediated by polyamine regulating protein, "antizyme."

2. What is antizyme?

Polyamines are highly charged bioactive substances presented ubiquitously in species from bacteria to human. Polyamines are necessary for cell growth and are involved in highly diversified cellular functions such as cell division, apoptosis, autophagy, oxidative stress, and ion channel activity. There are three major polyamines, putrescine, spermidine, and spermine, in the cells [5, 6]. Intracellular polyamine concentration is highly regulated by the protein "antizyme" [7–10] that is widely distributed from yeast to human [11]. Antizyme (AZ) is induced in response to the increased concentration of intracellular polyamines through the polyamine-induced translational frameshifting mechanism [12]. AZ mRNA consists of two ORFs (ORF1 and ORF2). In the low polyamine concentration, translation of ORF1 is terminated at stop codon "UGA" of ORF1, and short product is produced (**Figure 1**). But in the increasing cellular polyamine concentration, reading frame

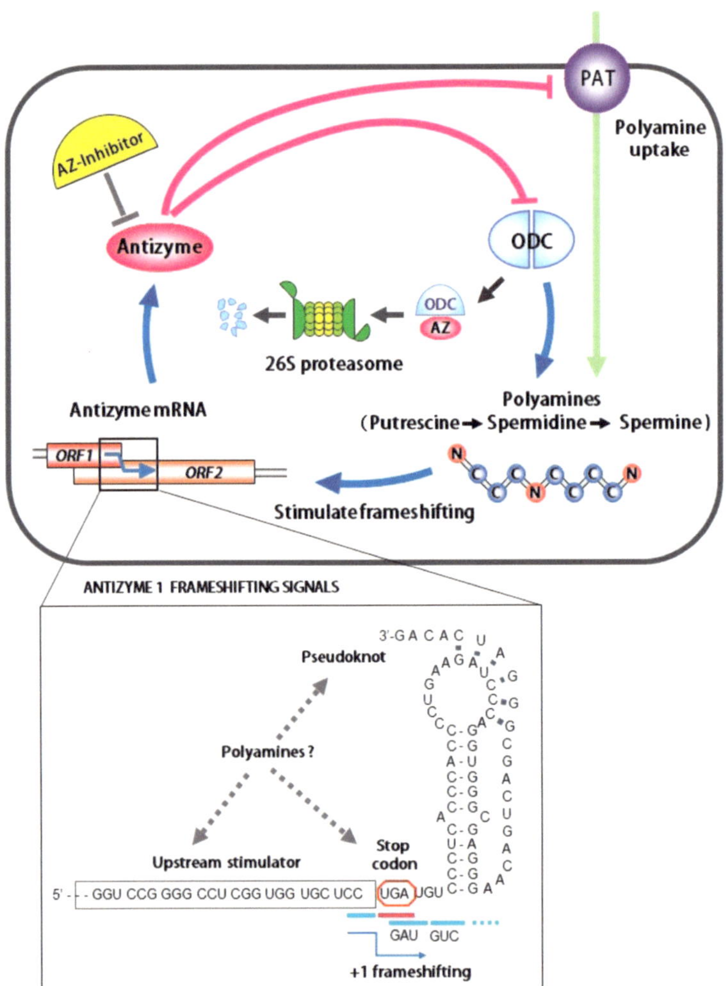

Figure 1.
Negative feedback regulation of cellular polyamine by antizyme. Three cis-acting elements, UGA stop codon, upstream stimulator, and pseudoknot structure, are known to be important for +1 frameshifting (bottom column). Putrescine, spermidine, and spermine are major polyamines in the mammalian cell. Putrescine synthesized from ornithine by ODC could be metabolized to spermidine and spermine in the cells. PAT is a polyamine transporter that uptakes polyamines from outside of the cells.

shifts +1 direction at the end of ORF1 (**Figure 1** bottom column). In this case, following ORF1, ORF2 is translated and full length active product "antizyme" is produced [12, 13]. The induced AZ protein binds to ornithine decarboxylase (ODC) monomer, a key enzyme in polyamine biosynthesis, and catalyzes the conversion from ornithine to putrescine and inhibits its activity. AZ-bound ODC is targeted to the 26S proteasome for degradation without ubiquitination (**Figure 1**) [14]. AZ also suppresses polyamine uptake by inhibiting membrane polyamine transporter (**Figure 1**) [15, 16]. Thus, AZ provides the negative feedback regulation of cellular polyamines. In addition, AZ is regulated by the protein, antizyme inhibitor (AZIN), that is homologous to ODC and can bind to AZ with higher affinity than ODC but lacking the enzymatic activity [17, 18].

In mammals, cells express three members of AZ protein family, AZ1–3 (**Table 1**) [19]. AZ1 and AZ2 are distributed ubiquitously in most of the tissues, whereas AZ3 is testis specific [20–22]. Both AZ1 and AZ2 bind to ODC and accelerate its

	AZ1	AZ2	AZ3
Species distribution	**Yeast-mammals**	**Vertebrates**	**Mammals**
Evolutional conservation	**Lower**	**Higher**	**Lower**
Tissue distribution	**Whole body**	**Whole body**	**Testis**
mRNA Expression	**High**	**Low**	**Low**
Cellular distribution	**Cytoplasm and Nucleus**	**Mainly nucleus**	**Spermatid and sperm**
Induction by polyamines	+	+	+
+1 frameshifting	+	+	+
ODC binding and inhibition	+	+	+
Acceleration of ODC degradation			
in vivo	+	+	-
in vitro	+	-	-
Inhibition of polyamine uptake	+	+	+
AZ inhibitor binding	+	+	+
Phenotype of knockout mice	**unreported**	**unreported**	**Male infertility**

Table 1.
Characteristics of antizyme family.

degradation in the cells [9, 23], but AZ3 has no activity for acceleration of ODC degradation [24]. The rate of ODC degradation by AZ1 is faster than that by AZ2 [23, 25]. Polyamine (putrescine) concentration of AZ1 knockdown cells is markedly increased, compared to that of AZ2 knockdown and control cells [26]. Therefore, it is thought that AZ1 mainly regulates cellular polyamine concentration. On the other hand, although AZ2 is highly homologous to AZ1 [25], it is considered that AZ2 is not a backup of AZ1 because of some differences between each other. AZ2 was found as one of the genes upregulated in neuronal cells by the drug that induces seizure [27]. Nucleic acid sequence of AZ2 is evolutionally conserved higher than that of AZ1 [11]. AZ2 is localized mainly in the nucleus [26] and is phosphorylated in the cells [28]. We will mention about AZ2 specific function with its interacting protein that we found very recently in this chapter.

3. Antizyme-interacting proteins and ubiquitin-independent proteasomal degradation

3.1 Antizyme 1-interacting proteins

It had been considered that ODC is the only protein degraded through AZ-mediated ubiquitin-independent proteasomal degradation system. However, recently several AZ1-interacting proteins other than ODC have been reported (**Table 2**). Although it has already been reported that those proteins are degraded by the ubiquitin-proteasome pathway, AZ1 could also accelerate those degradation without ubiquitination (**Tables 1** and **2**, **Figure 1**). Smad1, which is involved in bone morphogenetic protein (BMP) signaling pathway [29, 30], is the first reported protein that interacts AZ1 other than ODC [31]. In this case, newly synthesized HsN3, which is β-subunit for 20S proteasome, forms ternary complex with AZ1 and smad1. This complex may bind to 20S proteasome, and next 19S complex is docked on 20S, and then smad1 is degraded by the 26S proteasome.

Newman et al. reported that AZ1 has the ability to accelerate the degradation of cyclin D1, one of the cell cycle regulatory protein families [32]. Cyclin D1 interacts with cyclin-dependent kinase (CDK), and accumulation of cyclin D1-CDK complex is important for cell cycle progression [33]. This protein is already known to be degraded by ubiquitin-proteasome pathway [34]. They demonstrated that AZ1 induction by polyamine or overexpression of AZ1 accelerates cyclin D1 degradation, and knockdown of AZ1 suppresses it. Furthermore, in vitro experiment using purified cyclin D1, AZ1, and rabbit reticulocyte extracts as a source of 26S proteasome, AZ1 accelerated cyclin D1 degradation in a ATP-dependent manner. AZ1 could also degrade ubiquitin-deficient mutant of cyclin D1 in the cells [32]. In vitro size distribution analysis for binding between AZ1, cyclin D1, and ODC suggested that binding sites of AZ1 for cyclin D1 and ODC do not overlap each other, and cyclin D1 binds to the N-terminus of AZ1 and ODC binds at the C-terminus, respectively. Binding affinity of AZ1 to cyclin D1 is fourfold lower than that to ODC [35]. Although physiological significance is not clear, it showed that those three proteins form cyclin D1-AZ1-ODC ternary complex.

The oncogene Aurora A encodes a protein kinase that exerts essential roles in mitotic events and is important for induction of centrosome amplification [36]. Overexpression of Aurora A in many cancers induces aneuploidy, centrosome anomaly, poor prognosis, and invasiveness [37, 38]. Aurora A is ubiquitinated by the E3 ubiquitin (Ub) ligase, anaphase-promoting complex/cyclosome (APC/C) that is activated by both cell-division cycle protein 20 (Cdc20) and Cdh1,

Protein names	Protein function	Biological process	Characteristics of degradation
ODC	**Synthesis of putrescine from ornithinc**	**Polyamine metabolism**	**AZ1 or AZ2 binds to ODC monomer that is targeted to the 26S proteasome [14, 23]**
Smad1	**Binding to smad4 Regulation of transcription**	**BMP signaling pathway**	**Formation of HsN3-Smad1-AZ1 ternary complex [31].**
CyclinD1	**Binding to CDK**	**Cell cycle**	**AZ1 binds to Cyclin D1 and accelerates its degradation In vitro and in vivo [32].**
Aurora A	**Protein Kinase**	**Mitotic events**	**Formation of AUKAIP-1-AZ1-Aurora-A ternary complex, and that is target to the proteasome in vivo [41].**
Msp1	**Protein Kinase**	**Cell divition**	**AZ1 affects the level of Msp1 at the centrosome [46].**
DNp73	**Inhibition of both p73 and p53**	**Apoptosis**	**c-Jundependent DNp73 degradation mediated by AZ1 in vivo [52].**
c-Myc	**Transcriptio n factor**	**Cell growth Differentiat ion survival Apoptosis**	**AZ2 binds to c-Myc and accelerates its degradation in vivo [26].**

Table 2.
The proteins degraded by antizyme-mediated ubiquitin-independent proteasomal pathway.

substrate-recognition subunit of APC/C, and is degraded by the proteasome [39, 40]. However, Lim and Gopalan demonstrated that AZ1 could accelerate Aurora A degradation with ubiquitin-independent manner, where Aurora A kinase-interacting protein 1 (AURKAIP1), a negative regulator of Aurora A, enhances the binding of AZ1 to Aurora A and facilitates the recognition of Aurora A by the proteasome [41].

Mps1 is protein kinase required for centrosome duplication in regulating the spindle assembly checkpoint [42, 43]. Accumulation of Mps1 at the centrosome causes aberrant centriole assembly [44, 45]. In fact in various tumor cells, centrosomal Mps1 pool is increased, which causes abnormal centrosome duplication [44]. Thus degradation of Mps1 is important for proper pool of Mps1 at the centrosome. Although degradation of Mps1 is known to be mediated by the proteasome, amino acid residue 420–507 of the human Mps1 that is sufficient for its degradation does not contain APC/C recognition motifs, suggesting the commitment of Mps1 to ubiquitin-independent proteasomal degradation [45]. Kasbek et al. reported that AZ1 localizes to the centrosomes and binds to Mps1 to control the levels of centrosomal Mps1 by accelerating the degradation of Mps1 [46]. Fluorescent microscopy analysis showed that centrosomal Mps1 level is dependent on AZ1 expression, overexpression of AZ1 decreases the centrosome Mps1 level, and conversely, AZ1 knockdown by siRNA increases that. Furthermore, deletion of degradation signal of Mps1 abolished the regulation of centrosomal Mps1 level by AZ1. In addition, overexpressing AZIN in the cells to trap AZ1 and inhibit its function increased centrosomal Mps1 level. Thus the balance of AZ1 and Mps1 level in the centrosome is important for the centrosome duplication process.

P73 is a homolog of p53 and exists as two major forms, TAp73 or Delta-N (DN) p73. TAp73 is full-length form and exerts proapoptotic function, whereas DNp73, which is amino-terminal transactivation domain lacking the form of p73, exhibits dominant-negative inhibitor activity for both p73 and p53, resulting in antiapoptotic properties [47]. Therefore, in the stress condition like DNA damage, reduction of DNp73 level is needed to execute apoptosis [48–50]. It is known that degradation of both TAp73 and DNp73 is mediated by E3-ubiquitin ligase Itch in a proteasome-dependent manner in normal condition [51]. However, in Itch-decreased condition such as DNA damage by UV irradiation, stabilization of TAp73 was observed, but DNp73 was not [51]. Therefore, it was considered that the degradation of TAp73 and DNp73 is regulated by different mechanisms. Dulloo et al. reported that reduction of DNp73 in the stress condition is due to the degradation of DNp73 by AZ1-mediated ubiquitin-independent proteasomal pathway [52]. They showed that degradation of DNp73 could be induced by genotoxic stresses such as UV irradiation and doxorubicin treatment. Inhibition of ubiquitin-activating enzyme E1 by the inhibitor PYR41 could not block DNp73 degradation, indicating that it relies on ubiquitin-independent pathway. They demonstrated that polyamine induced AZ1 to bind to DNp73 for accelerating its degradation. Interestingly, AZ1-mediated DNp73 degradation is dependent on transcription factor c-Jun that is activated by stress signals. Overexpression and knocking down of AZ1 also showed that even in the presence of c-Jun, AZ1 is necessary for genotoxic stress to induce DNp73 degradation. Although it is not clear how c-Jun operates AZ1 expression, c-Jun may act upstream of polyamine biosynthesis pathway.

Thus, several proteins degraded by AZ1-mediated proteasome pathway are found, but AZ2-interacting protein or AZ2-mediated proteasomal degradation other than ODC has not been reported. We recently found two AZ2-interacting proteins, and one of the two was the protein that accelerated its proteasomal degradation by AZ2 without ubiquitination (see next section).

3.2 Antizyme 2-interacting proteins

As mentioned above, AZ2 also binds to ODC and accelerates its degradation in the cells [9]. However, we have considered that AZ2 has specific function other than AZ1 because of the differences such as nuclear localization [26, 28], highly gene conservation between species [20], and high expression in neuronal cells [53].

We performed comprehensive analysis of AZ2-interacting protein using two-hybrid technique. Two AZ2-interacting proteins were identified. One is ATP citrate lyase (ACLY), which is the enzyme catalyzing acetyl-CoA production in cytosol [54] and related to lipid anabolism and acetylation of cellular components [55]. We found that ACLY binds not only to AZ2 but also to AZ1 by immunoprecipitation assay [56]. Degradation assay for ACLY was performed in expectation of ubiquitin-independent proteasomal degradation. However, AZs have no ability to accelerate ACLY degradation. Surprisingly, AZ1 and AZ2 activate catalytic activity of ACLY [56]. The other is proto-oncogene c-Myc that is a transcription factor with a basic region/helix–loop–helix/leucine zipper domain and forms heterodimer with Max for DNA binding [57, 58]. c-Myc functions as a master regulator of a variety of cellular processes such as cell growth, differentiation, survival, and apoptosis [58]. In cell growth, c-Myc targets ODC gene [59] and promotes synthesis of polyamine that is important for stabilization of nucleic acids, transcription, translation, and +1 frameshifting on AZ mRNA [6].

It is known that degradation of c-Myc is mediated by ubiquitin-proteasome pathway, where c-Myc is phosphorylated at Thr-58 (pT58) and Ser-62 (pS62) by extracellular signal-regulated kinase, ERK, and glycogen synthase kinase 3β, GSK-3β, respectively [60, 61]. After dephosphorylation at Ser-62 by protein phosphatase 2A, PP2A, pT58-c-Myc is ubiquitinated by E3-ubiquitin ligase Fbxw7 for proteasomal degradation [60, 62]. At first, AZ2-interacting protein identified by the comprehensive analysis mentioned above was not c-Myc but a protein that has basic region/helix–loop–helix/leucine zipper domain and interacts with c-Myc (Murai et al., manuscript in preparation). However, in the process of analyzing the interaction with AZ2, we found that AZ2 interacts with c-Myc in the cells by immunoprecipitation assay. Subcellular localization analysis of both proteins using fluorescent protein tags or antibody conjugated fluorescent probe revealed that AZ2 co-localized with

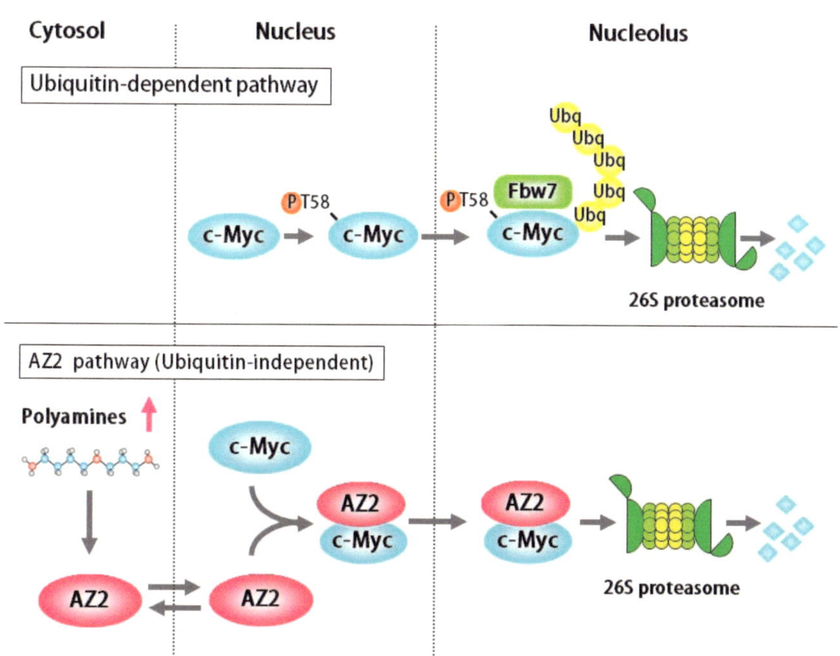

Figure 2.
AZ2-mediated c-MYC degradation in the nucleolus. Two distinct c-Myc degradation pathways exist in the cells. It is thought that AZ2 pathway functions under the stress condition (polyamine increased condition) such as glucose-free and hypoxia.

c-Myc in the nucleus. Treatment of proteasome inhibitor MG132 changes the nuclear co-localization of both proteins to nucleolar co-localization [26]. Overexpression of AZ2 or addition of polyamine in the cells accelerated c-Myc degradation, and knock-down of AZ2 with siRNA suppressed it. Furthermore, E1 inhibitor PYR-41 could not suppress AZ2-mediated proteasomal c-Myc degradation [26]. These results suggest that AZ2-mediated ubiquitin-independent nucleolar c-Myc degradation pathway other than ubiquitin-dependent one exists in the cells (**Figure 2**).

4. Conclusions

In this chapter, antizyme-mediated ubiquitin-independent proteasomal degradation has been discussed. All the proteins mentioned above are already known as the proteins degraded by ubiquitin-proteasomal pathway. It is not clear how antizyme-mediated ubiquitin-independent degradation of these proteins is physiologically significant. Normally subcellular localization of ODC is mainly in the cytoplasm and at least not in the nucleolus even in the presence of MG132. In addition, ODC is necessary for cell growth, and the affinity of interaction between antizyme and ODC is high [63]; in such condition, ODC probably occupies almost all antizymes in the cytosol, and antizymes hardly function for other antizyme-interacting proteins [64]. In this context, because subcellular localization of both AZ2 and its interacting protein c-Myc is in the nucleus or nucleolus, cytosolic protein ODC could not interact with AZ2 there. ODC is one of the c-Myc-targeting proteins, and AZ2 may function upstream of c-Myc especially under the stress condition such as glucose free and hypoxic condition [26]. Further studies are needed to elucidate the significance of antizyme-proteasome degradation pathway.

Acknowledgements

This research was supported by the Jikei University Graduate Research Fund and JSPS KAKENHI Grant Number JP 19K08283.

Author details

Noriyuki Murai
Department of Molecular Biology, The Jikei University School of Medicine, Tokyo, Japan

*Address all correspondence to: nmurai@jikei.ac.jp

References

[1] Hershko A, Ciechanover A. The ubiquitin system. Annual Review of Biochemistry. 1998;**67**:425-479

[2] Glickman MH, Ciechanover A. The ubiquitin-proteasome protcolytic pathway: Destruction for the sake of construction. Physiological Reviews. 2002;**82**(2):373-428

[3] Jariel-Encontre I, Bossis G, Piechaczyk M. Ubiquitin-independent degradation of proteins by the proteasome. Biochimica et Biophysica Acta. 2008;**1786**(2):153-177

[4] Buneeva OA, Medvedev AE. Ubiquitin-independent protein degradation in proteasomes. Bio-meditsinskaya Khimiya. 2018;**64**(2): 134-148

[5] Pegg AE, Casero RA Jr. Current status of the polyamine research field. Methods in Molecular Biology (Clifton, NJ). 2011;**720**:3-35

[6] Casero RA Jr, Murray Stewart T, Pegg AE. Polyamine metabolism and cancer: Treatments, challenges and opportunities. Nature Reviews. Cancer. 2018;**18**(11):681-695

[7] Hayashi S, Murakami Y, Matsufuji S. Ornithine decarboxylase antizyme: A novel type of regulatory protein. Trends in Biochemical Sciences. 1996;**21**(1): 27-30

[8] Li X, Coffino P. Regulated degradation of ornithine decarboxylase requires interaction with the polyamine-inducible protein antizyme. Molecular and Cellular Biology. 1992;**12**(8):3556-3562

[9] Coffino P. Regulation of cellular polyamines by antizyme. Nature Reviews. Molecular Cell Biology. 2001;**2**(3):188-194

[10] Heller JS, Fong WF, Canellakis ES. Induction of a protein inhibitor to ornithine decarboxylase by the end products of its reaction. Proceedings of the National Academy of Sciences of the United States of America. 1976;**73**(6):1858-1862

[11] Ivanov IP, Gesteland RF, Atkins JF. Antizyme expression: A subversion of triplet decoding, which is remarkably conserved by evolution, is a sensor for an autoregulatory circuit. Nucleic Acids Research. 2000;**28**(17):3185-3196

[12] Matsufuji S, Matsufuji T, Miyazaki Y, Murakami Y, Atkins JF, Gesteland RF, et al. Autoregulatory frameshifting in decoding mammalian ornithine decarboxylase antizyme. Cell. 1995;**80**(1):51-60

[13] Atkins JF, Loughran G, Bhatt PR, Firth AE, Baranov PV. Ribosomal frameshifting and transcriptional slippage: From genetic steganography and cryptography to adventitious use. Nucleic Acids Research. 2016;**44**(15): 7007-7078

[14] Murakami Y, Matsufuji S, Kameji T, Hayashi S, Igarashi K, Tamura T, et al. Ornithine decarboxylase is degraded by the 26S proteasome without ubiquitination. Nature. 1992;**360**(6404):597-599

[15] Mitchell JL, Judd GG, Bareyal-Leyser A, Ling SY. Feedback repression of polyamine transport is mediated by antizyme in mammalian tissue-culture cells. The Biochemical Journal. 1994;**299**(Pt 1):19-22

[16] Suzuki T, He Y, Kashiwagi K, Murakami Y, Hayashi S, Igarashi K. Antizyme protects against abnormal accumulation and toxicity of polyamines in ornithine decarboxylase-overproducing cells. Proceedings of the National Academy of Sciences of the United States of America. 1994;**91**(19):8930-8934

[17] Fujita K, Murakami Y, Hayashi S. A macromolecular inhibitor of the antizyme to ornithine decarboxylase. The Biochemical Journal. 1982;**204**(3):647-652

[18] Kitani T, Fujisawa H. Purification and characterization of antizyme inhibitor of ornithine decarboxylase from rat liver. Biochimica et Biophysica Acta. 1989;**991**(1):44-49

[19] Ivanov IP, Gesteland RF, Atkins JF. Survey and summary: Antizyme expression: A subversion of triplet decoding, which is remarkably conserved by evolution, is a sensor for an autoregulatory circuit. Nucleic Acids Research. 2000;**28**(17):3185-3196

[20] Ivanov IP, Gesteland RF, Atkins JF. A second mammalian antizyme: Conservation of programmed ribosomal frameshifting. Genomics. 1998;**52**(2):119-129

[21] Ivanov IP, Rohrwasser A, Terreros DA, Gesteland RF, Atkins JF. Discovery of a spermatogenesis stage-specific ornithine decarboxylase antizyme: Antizyme 3. Proceedings of the National Academy of Sciences of the United States of America. 2000;**97**(9):4808-4813

[22] Tosaka Y, Tanaka H, Yano Y, Masai K, Nozaki M, Yomogida K, et al. Identification and characterization of testis specific ornithine decarboxylase antizyme (OAZ-t) gene: Expression in haploid germ cells and polyamine-induced frameshifting. Genes to Cells: Devoted to Molecular & Cellular Mechanisms. 2000;**5**(4):265-276

[23] Chen H, MacDonald A, Coffino P. Structural elements of antizymes 1 and 2 are required for proteasomal degradation of ornithine decarboxylase. The Journal of Biological Chemistry. 2002;**277**(48):45957-45961

[24] Snapir Z, Keren-Paz A, Bercovich Z, Kahana C. Antizyme 3 inhibits polyamine uptake and ornithine decarboxylase (ODC) activity, but does not stimulate ODC degradation. The Biochemical Journal. 2009;**419**(1):99-103. 1 p following

[25] Zhu C, Lang DW, Coffino P. Antizyme 2 is a negative regulator of ornithine decarboxylase and polyamine transport. The Journal of Biological Chemistry. 1999;**274**(37):26425-26430

[26] Murai N, Murakami Y, Tajima A, Matsufuji S. Novel ubiquitin-independent nucleolar c-Myc degradation pathway mediated by antizyme 2. Scientific Reports. 2018;**8**(1):3005

[27] Kajiwara K, Nagawawa H, Shimizu-Nishikawa S, Ookuri T, Kimura M, Sugaya E. Molecular characterization of seizure-related genes isolated by differential screening. Biochemical and Biophysical Research Communications. 1996;**219**(3):795-799

[28] Murai N, Shimizu A, Murakami Y, Matsufuji S. Subcellular localization and phosphorylation of antizyme 2. Journal of Cellular Biochemistry. 2009;**108**(4):1012-1021

[29] Kawabata M, Imamura T, Miyazono K. Signal transduction by bone morphogenetic proteins. Cytokine & Growth Factor Reviews. 1998;**9**(1):49-61

[30] Bragdon B, Moseychuk O, Saldanha S, King D, Julian J, Nohe A. Bone morphogenetic proteins: A critical review. Cellular Signalling. 2011;**23**(4):609-620

[31] Lin Y, Martin J, Gruendler C, Farley J, Meng X, Li BY, et al. A novel link between the proteasome pathway and the signal transduction pathway of the bone morphogenetic proteins (BMPs). BMC Cell Biology. 2002;**3**:15

[32] Newman RM, Mobascher A, Mangold U, Koike C, Diah S,

Schmidt M, et al. Antizyme targets cyclin D1 for degradation. A novel mechanism for cell growth repression. The Journal of Biological Chemistry. 2004;**279**(40):41504-41511

[33] Diehl JA. Cycling to cancer with cyclin D1. Cancer Biology & Therapy. 2002;**1**(3):226-231

[34] Diehl JA, Zindy F, Sherr CJ. Inhibition of cyclin D1 phosphorylation on threonine-286 prevents its rapid degradation via the ubiquitin-proteasome pathway. Genes & Development. 1997;**11**(8):957-972

[35] Liu YC, Lee CY, Lin CL, Chen HY, Liu GY, Hung HC. Multifaceted interactions and regulation between antizyme and its interacting proteins cyclin D1, ornithine decarboxylase and antizyme inhibitor. Oncotarget. 2015;**6**(27):23917-23929

[36] Bischoff JR, Anderson L, Zhu Y, Mossie K, Ng L, Souza B, et al. A homologue of *Drosophila aurora* kinase is oncogenic and amplified in human colorectal cancers. The EMBO Journal. 1998;**17**(11):3052-3065

[37] Sakakura C, Hagiwara A, Yasuoka R, Fujita Y, Nakanishi M, Masuda K, et al. Tumour-amplified kinase BTAK is amplified and overexpressed in gastric cancers with possible involvement in aneuploid formation. British Journal of Cancer. 2001;**84**(6):824-831

[38] Buschhorn HM, Klein RR, Chambers SM, Hardy MC, Green S, Bearss D, et al. Aurora-A over-expression in high-grade PIN lesions and prostate cancer. The Prostate. 2005;**64**(4):341-346

[39] Littlepage LE, Ruderman JV. Identification of a new APC/C recognition domain, the A box, which is required for the Cdh1-dependent destruction of the kinase Aurora-A during mitotic exit. Genes & Development. 2002;**16**(17):2274-2285

[40] Taguchi S, Honda K, Sugiura K, Yamaguchi A, Furukawa K, Urano T. Degradation of human Aurora-A protein kinase is mediated by hCdh1. FEBS Letters. 2002;**519**(1-3):59-65

[41] Lim SK, Gopalan G. Antizyme1 mediates AURKAIP1-dependent degradation of Aurora-A. Oncogene. 2007;**26**(46):6593-6603

[42] Abrieu A, Magnaghi-Jaulin L, Kahana JA, Peter M, Castro A, Vigneron S, et al. Mps1 is a kinetochore-associated kinase essential for the vertebrate mitotic checkpoint. Cell. 2001;**106**(1):83-93

[43] Stucke VM, Silljé HH, Arnaud L, Nigg EA. Human Mps1 kinase is required for the spindle assembly checkpoint but not for centrosome duplication. The EMBO Journal. 2002;**21**(7):1723-1732

[44] Kasbek C, Yang CH, Fisk HA. Mps1 as a link between centrosomes and genomic instability. Environmental and Molecular Mutagenesis. 2009;**50**(8):654-665

[45] Kasbek C, Yang CH, Yusof AM, Chapman HM, Winey M, Fisk HA. Preventing the degradation of mps1 at centrosomes is sufficient to cause centrosome reduplication in human cells. Molecular Biology of the Cell. 2007;**18**(11):4457-4469

[46] Kasbek C, Yang CH, Fisk HA. Antizyme restrains centrosome amplification by regulating the accumulation of Mps1 at centrosomes. Molecular Biology of the Cell. 2010;**21**(22):3878-3889

[47] Melino G, De Laurenzi V, Vousden KH. p73: Friend or foe in tumorigenesis. Nature Reviews. Cancer. 2002;**2**(8):605-615

[48] Irwin MS, Kondo K, Marin MC, Cheng LS, Hahn WC, Kaelin WG Jr. Chemosensitivity linked to p73 function. Cancer Cell. 2003;**3**(4):403-410

[49] Lin KW, Nam SY, Toh WH, Dulloo I, Sabapathy K. Multiple stress signals induce p73beta accumulation. Neoplasia (New York, NY). 2004;**6**(5):546-557

[50] Maisse C, Munarriz E, Barcaroli D, Melino G, De Laurenzi V. DNA damage induces the rapid and selective degradation of the DeltaNp73 isoform, allowing apoptosis to occur. Cell Death and Differentiation. 2004;**11**(6):685-687

[51] Rossi M, De Laurenzi V, Munarriz E, Green DR, Liu YC, Vousden KH, et al. The ubiquitin-protein ligase Itch regulates p73 stability. The EMBO Journal. 2005;**24**(4):836-848

[52] Dulloo I, Gopalan G, Melino G, Sabapathy K. The antiapoptotic DeltaNp73 is degraded in a c-Jun-dependent manner upon genotoxic stress through the antizyme-mediated pathway. Proceedings of the National Academy of Sciences of the United States of America. 2010;**107**(11):4902-4907

[53] Ramos-Molina B, Lopez-Contreras AJ, Cremades A, Penafiel R. Differential expression of ornithine decarboxylase antizyme inhibitors and antizymes in rodent tissues and human cell lines. Amino Acids. 2012;**42**(2-3):539-547

[54] Watson JA, Fang M, Lowenstein JM. Tricarballylate and hydroxycitrate: Substrate and inhibitor of ATP: Citrate oxaloacetate lyase. Archives of Biochemistry and Biophysics. 1969;**135**(1):209-217

[55] Chypre M, Zaidi N, Smans K. ATP-citrate lyase: A mini-review. Biochemical and Biophysical Research Communications. 2012;**422**(1):1-4

[56] Tajima A, Murai N, Murakami Y, Iwamoto T, Migita T, Matsufuji S. Polyamine regulating protein antizyme binds to ATP citrate lyase to accelerate acetyl-CoA production in cancer cells. Biochemical and Biophysical Research Communications. 2016

[57] Meyer N, Penn LZ. Reflecting on 25 years with MYC. Nature Reviews. Cancer. 2008;**8**(12):976-990

[58] van Riggelen J, Yetil A, Felsher DW. MYC as a regulator of ribosome biogenesis and protein synthesis. Nature Reviews. Cancer. 2010;**10**(4):301-309

[59] Bello-Fernandez C, Packham G, Cleveland JL. The ornithine decarboxylase gene is a transcriptional target of c-Myc. Proceedings of the National Academy of Sciences of the United States of America. 1993;**90**(16):7804-7808

[60] Yada M, Hatakeyama S, Kamura T, Nishiyama M, Tsunematsu R, Imaki H, et al. Phosphorylation-dependent degradation of c-Myc is mediated by the F-box protein Fbw7. The EMBO Journal. 2004;**23**(10):2116-2125

[61] Welcker M, Orian A, Jin J, Grim JE, Harper JW, Eisenman RN, et al. The Fbw7 tumor suppressor regulates glycogen synthase kinase 3 phosphorylation-dependent c-Myc protein degradation. Proceedings of the National Academy of Sciences of the United States of America. 2004;**101**(24):9085-9090

[62] Yeh E, Cunningham M, Arnold H, Chasse D, Monteith T, Ivaldi G, et al. A signalling pathway controlling c-Myc degradation that impacts oncogenic transformation of human cells. Nature Cell Biology. 2004;**6**(4):308-318

[63] Liu YC, Hsu DH, Huang CL, Liu YL, Liu GY, Hung HC. Determinants of the differential antizyme-binding affinity

of ornithine decarboxylase. PLOS One. 2011;**6**(11):e26835

[64] Bercovich Z, Snapir Z, Keren-Paz A, Kahana C. Antizyme affects cell proliferation and viability solely through regulating cellular polyamines. The Journal of Biological Chemistry. 2011;**286**(39):33778-33783

Chapter 3

Lys63-Linked Polyubiquitination of Transforming Growth Factor β Type I Receptor (TβRI) Specifies Oncogenic Signaling

Jie Song and Maréne Landström

Abstract

Transforming growth factor β (TGFβ) is a multifunctional cytokine with potent regulatory effects on cell fate during embryogenesis, in the normal adult organism, and in cancer cells. In normal cells, the signal from the TGFβ ligand is transduced from the extracellular space to the cell nucleus by transmembrane serine–threonine kinase receptors in a highly specific manner. The dimeric ligand binding to the TGFβ Type II receptor (TβRII) initiates the signal and then recruits the TGFβ Type I receptor (TβRI) into the complex, which activates TβRI. This causes phosphorylation of receptor-activated Smad proteins Smad2 and Smad3 and promotes their nuclear translocation and transcriptional activity in complex with context-dependent transcription factors. In several of our most common forms of cancer, this pathway is instead regulated by polyubiquitination of TβRI by the E3 ubiquitin ligase TRAF6, which is associated with TβRI. The activation of TRAF6 promotes the proteolytic cleavage of TβRI, liberating its intracellular domain (TβRI-ICD). TβRI-ICD enters the cancer cell nucleus in a manner dependent on the endosomal adaptor proteins APPL1/APPL2. Nuclear TβRI-ICD promotes invasion by cancer cells and is recognized as acting distinctly and differently from the canonical TGFβ-Smad signaling pathway occurring in normal cells.

Keywords: TRAF6, APPL1/2, TGFβ, biomarkers, cancer

1. Introduction

Ubiquitination is a crucial biological process both in normal homeostasis and in diseases including cancer and immunity-related disorders. In cancers, ubiquitination of various signaling molecules acts to both promote and suppress tumors [1]. In this chapter, we will focus on the tumor-promoting role of TRAF6 in different cancers.

1.1 Ubiquitination and TRAF6

Within the lifespan of proteins, it is difficult for them to avoid post-translational modifications, which determine their localization and function. Protein ubiquitination was discovered in the early 1980s, and is a dynamic post-translational modification regulating many cellular processes. The best known role for ubiquitination is targeting proteins for their destruction by the proteasome. In recent years, however, nonproteolytic functions of ubiquitination, including in signal transduction, cell division and differentiation, endocytosis, and the DNA damage response, have been rapidly discovered [2].

Ubiquitin is a highly conserved protein of 76 amino acids that becomes covalently attached to the ε-amino group of lysine (Lys) residues of target proteins. There are three types of ubiquitination: mono-ubiquitination, multi-mono-ubiquitination, and polyubiquitination. Polyubiquitin chains are formed by the addition of ubiquitin residues to an ubiquitin molecule already linked to a protein and acting as an additional residue. The key features of ubiquitin are seven Lys residues (Lys6, Lys11, Lys27, Lys29, Lys33, Lys48, and Lys63) and an N-terminal Met residue, all of which can be further ubiquitinated to give rise to polyubiquitin chains of distinct linkages. Lys63-linked polyubiquitination is involved in endocytosis, signal transduction, and DNA-damage tolerance [3, 4]. During recent years has also linear ubiquitination been identified to occur through N-terminal Met residue of ubiquitin. It is created by the linear ubiquitin chain assembly complex (LUBAC), which so far is the only ubiquitin ligase known to build linear ubiquitin chains *de novo*. Linear ubiquitination is crucial for regulation of innate and adaptive immune responses, including inflammatory responses and regulation of signals leading to cell death [5–7].

Ubiquitination is catalyzed by a sophisticated enzymatic cascade involving three enzymes, an ubiquitin-activating enzyme (E1), an ubiquitin-conjugating enzyme (E2), and an ubiquitin ligase (E3). E3 ligase usually determines the mechanism of ubiquitin transfer to target proteins, as it can recognize substrate and mediate the addition of ubiquitin [3, 8]. E3 ligases have been classified into three subfamilies: HECT (homologous to E6-AP C terminus) ligases, RING (really interesting new gene)/U-box ligases, and RBR (RING-between-RING) ligases [3]. TRAF6 is a Ring/U-box E3 ligase belonging to the tumor necrosis factor (TNF) receptor-associated factor (TRAF) family.

TRAF family cytoplasmic proteins were originally identified as TNF receptor signaling adaptors that bind directly to the cytoplasmic region of TNF-R2. To date, six different TRAF family members (TRAF1–6) have been found in mammals. TRAF7 is controversially classified as a member of the TRAF family, as it lacks a TRAF homology domain and does not directly bind to any member of the TNFR superfamily, two key features used to define the TRAF family. The TRAF domain, located in the C-terminal portion of TRAF family proteins, is composed of an N-terminal less-conserved coiled-coil region (TRAF-N) and a C-terminal highly conserved subdomain (TRAF-C). The TRAF domain mediates protein–protein interactions, including association with upstream regulators and downstream effectors and homo- and hetero-dimerization of TRAF proteins. Thus, TRAF family members are involved in a variety of signal transduction pathways by interaction with receptors. These include the TNF, Toll-like receptor, NLR, TGFβ signaling pathways, and others. Through these interactions, TRAF family members participate in the regulation of a broad range of cellular processes, including proliferation, differentiation, apoptosis, and survival. With the exception of TRAF1, however, TRAFs also contain an N-terminal RING domain, indicating that they are E3 ubiquitin ligases [9, 10].

TRAF6 was isolated for the first time in 1996 in a yeast two-hybrid screen with CD40 as bait [11], and later independently found to mediate the expression of interleukin 1 (IL-1) signaling, based on a screen of an EST expression library [12]. TRAF6 is well conserved across species and broadly expressed in mammalian tissues such as brain, lung, liver, etc. As an E3 ligase, TRAF6 interacts with the E2 complex Ubc13-Uev1A and participates in a number of signal transduction pathways, including those of nuclear factor kappa B (NF-κB), toll-like receptor 4 (TLR4), and TGFβ, the last of which is further discussed later in this chapter. Knockdown of TRAF6 or inhibition of TRAF6 E3 ligase activity *in vitro* suppresses the proliferation, survival, migration, invasion, and metastasis of many human epithelial cell lines [10].

TRAF6$^{-/-}$ mice, with a complete lack of normal T and B cell areas, exhibit perinatal or postnatal death due to severe splenomegaly, osteopetrosis, lymph node deficiency, and thymic atrophy [9]. All these findings indicate the critical and highly various roles of TRAF6 in cytokine signaling, innate immune responses, and perinatal and postnatal survival [9, 13].

1.2 The TGFβ signaling pathway and its role in cancer

Cells communicate by sending and receiving signals through cytokines and membrane-associated proteins. Among the secreted growth factors and cytokines, the TGFβ family attracts a lot of attention because it controls cell fate decisions during embryonic development, tissue homeostasis, and regeneration. All cells in the developing embryo and the adult can respond to TGFβ, as it regulates proliferation, differentiation, adhesion, movement, and apoptosis in a cell-context–dependent manner. Perturbation of TGFβ signaling is often seen in inflammatory diseases, fibrotic diseases, and cancers [14, 15].

1.2.1 Basics of TGFβ signaling

The TGFβ superfamily consists of more than 30 members in humans, and they are grouped into different subfamilies based on sequence similarity and functional criteria, including TGFβ isoforms, activins, bone morphogenetic proteins (BMPs), growth and differentiation factors (GDFs), activin, nodal, and anti-mullerian hormone (AMH). The TGFβ subfamily comprises three different isoforms: TGFβ1, TGFβ2, and TGFβ3. All of them act in an autocrine, paracrine, and sometimes endocrine manner [14, 16].

Mammalian genomes encode two subfamilies of TGFβ receptors, seven type I (TβRI) and five type II (TβRII) serine/threonine kinase receptors, which are classified by their structures and functions. Both types of receptors are single-pass transmembrane kinases and share structural similarities: they have an N-terminal cysteine-rich extracellular domain, an α-helical transmembrane domain, a short juxtamembrane sequence, and a C-terminal cytoplasmic kinase domain with 11 subdomains organized in an N-lobe and a C-lobe. A conserved glycine/serine-rich sequence, the GS domain, is present in the juxtamembrane domain only in TβRI [17, 18].

The most-studied mediators of TGFβ signaling pathways are Smad proteins. TGFβ signaling pathways include canonical Smad-dependent and non-canonical Smad- independent pathways [15, 19].

1.2.2 Smad-dependent TGFβ signaling pathways

Smad proteins are named after two proteins: *small body size (Sma)* in *Caenorhabditis elegans* and *mothers against decapentaplegic (Mad)* in *Drosophila melanogaster*. The mammalian genome encodes eight Smads which form three subfamilies based on their structures and functions: receptor-activated Smads (R-Smads; Smad 1, 2, 3, 5, and 8); a single common mediator of Smad (Co-Smad; Smad4); and two inhibitory Smads (I-Smads; Smad6 and Smad7). Smad2 and

Smad3 act as signal transducers for TGFβ, activin, and nodals, whereas Smad1, Smad5, and Smad8 mediate signals by BMPs and GDFs.

Upon TGFβ ligand binding, the two types of receptors are brought together and induce the formation of a heterotetrameric complex. The constitutively active type II receptor phosphorylates the type I receptor in its highly conserved GS domain, leading to the activation of its kinase. The active serine/threonine type I receptor propagates signaling by phosphorylating R-Smads, which in turn form a trimeric complex with Smad4 and then translocate to the nucleus. In the nucleus, the Smad complex works together with other transcription factors, coactivators, and corepressors to regulate the expression of genes such as snail family transcriptional repressor 1 (Snail1), plasminogen activator inhibitor type 1(PAI1), and matrix metallopeptidase 2 (MMP2). In summary, canonical Smad-dependent TGFβ signaling pathways regulate cell proliferation, apoptosis, and the epithelial-mesenchymal transition (EMT) [20, 21].

1.2.3 Smad-independent TGFβ signaling pathways

TGFβ non-canonical signaling pathways include the c-Jun N-terminal kinase (JNK), p38 mitogen-activated protein kinase (MAPK), phosphatidylinositol-3′-kinase (PI3K), and extracellular signal-regulated kinase (Erk) signaling pathways [19].

TGFβ-activated kinase-1 (TAK1) is a serine/threonine kinase and member of the mitogen-activated protein kinase (MAPK) kinase kinase (MAPKKK) family. TRAF6 associates with a conserved consensus motif in TβRI. Upon TGFβ stimulation, the interaction of TRAF6 and TβRI is important for the autoubiquitination of TRAF6 and subsequent Lys63-polyubiquitination and activation of TAK1. Once activated, TAK1 phosphorylates protein mitogen-activated kinase kinase 3/6 (MKK3/6), activating the JNK and p38 signaling pathways to drive apoptosis or EMT [22, 23].

1.2.4 TGFβ receptor endocytosis

Endocytosis is a process by which cells internalize extracellular materials and portions of their plasma membrane together with cell surface receptors. It has been divided into two categories, clathrin-dependent and clathrin-independent endocytosis [24]. TGFβRs can be internalized via both clathrin-dependent and clathrin-independent caveolae-mediated endocytosis [14, 25].

Both TβRII and TβRI appear to undergo rapid internalization in the presence and absence of ligand stimulation. After internalization, TGFβRs are found in the phosphatidylinositol-3-phosphate (PI3P)-enriched and early endosome antigen (EEA1)-positive endosomes, which recruit Smad anchor for receptor activation (SARA) to facilitate phosphorylation of R-Smads. Phosphorylated R-Smads in endosomes then dissociate from SARA and the receptors, and translocate to the nucleus together with Smad4 to regulate target gene expression [26].

In caveolae-mediated endocytosis, TGFβ signaling is turned off by the interaction between TGFβRs and Smad7-Smurf2, which leads to the degradation of TGFβRs [27].

1.2.5 TGFβ signaling in cancer

TGFβ signaling in cancer is a double-edged sword, acting as both a tumor suppressor in normal and pre-malignant cells and as a tumor promoter in malignant cells. The response to TGFβ is context dependent. TGFβ is produced by cancer cells or stromal cells in large amounts within the cancer microenvironment, influencing not only on the cancer cells but also non-tumor cells, such as fibroblasts and immune cells [15, 28].

In the early malignant stage, TGFβ suppresses tumor progression by inducing apoptosis and inhibiting proliferation. However, malignant cells always escape this tumor-suppressive response through loss of the core TGFβ pathway or its suppressive arms, thereby turning TGFβ into a stimulator of cancer progression. As a tumor promoter, TGFβ is involved in angiogenesis, tumor growth, evasion of immune surveillance, migration, invasion, and metastasis [15, 29].

1.3 PI3K/AKT pathway

The PI3K pathway is one of the most commonly activated pathways in human cancers, regulating cell proliferation, survival, metabolism, and vesicle trafficking. This pathway's activation is initiated by various molecules, such as insulin, glucose, growth factors, and cytokines [30, 31]. PI3Ks are classified into three classes based on sequence homology and substrate specificity. Class I PI3Ks have two subfamilies, IA and IB, classified according to their different regulatory mechanisms. Class IA PI3K, a heterodimer, consists of a p110 catalytic subunit and a p85 regulatory subunit. Class I PI3K generates PtdIns $[3,4,5]P_3$ (PIP_3) from PtdIns $[4,5]P_2$ (PIP_2) *in vivo*. PIP_3 acts as a second messenger to activate downstream signaling pathways, including AKT/mTOR (mechanistic target of rapamycin kinase) pathways. Class IA PI3Ks are the focus of this chapter [31, 32].

The primary structure of p85 includes an N-terminal Src homology 3 (SH3) domain, a RhoGap homology region located between two proline-rich domains, and two SH2 domains (nSH2 and cSH2 domains) separated by a p110-binding iSH2 domain [33]. Upon binding to an activated receptor tyrosine kinase (RTK) or G protein-coupled receptor (GPCP), p85 interacts with receptors directly or indirectly via the SH2 domains, which mediate the translocation of the p85-p110 complex to the cell membrane. This induces a conformational change and activates the catalytic activity of p110 to phosphorylate PIP_2 to generate PIP_3 [30, 33].

The serine/threonine protein kinase AKT has three isoforms, AKT1, AKT2, and AKT3. PIP_3 binding induces a conformational change in AKT that recruits phosphoinositide-dependent kinase (PDK1) to phosphorylate AKT on Thr308. The mTOR complex 2 (mTORC2) phosphorylates AKT on Ser473, fully activating AKT. All three isoforms are activated in the same manner [31, 34]. In addition to phosphorylation, other post-translational modifications regulate the activity of AKT. These include dephosphorylation, glycosylation, acetylation, ubiquitination, and SUMOylation. Lys48-linked polyubiquitination of AKT is mediated by multiple E3 ligases, such as BRCA1, Chip, MULAN, and TTC, and has been shown to promote proteasome-dependent degradation. By contrast, Lys63-linked polyubiquitination, which is mediated by TRAF6, Skp2, and NEDD4, is implicated in the membrane localization and phosphorylation of AKT [34, 35]. After activation, AKT regulates downstream signaling pathways by phosphorylating protein targets, including protein kinases, transcription factors, metabolic enzymes, cell cycle proteins, and others [34].

It has been reported that TGFβ can activate the PI3K signaling pathway directly or indirectly. Of note, upon TGFβ stimulation, the phosphorylation of AKT acts in a Smad-independent manner [36–38]. Moreover, p85 constitutively interacts with TβRII and binds to TβRI after TGFβ stimulation [39]. The crosstalk between the PI3K/AKT and TGFβ signaling pathways attracts a lot of attention, as both of them play important roles in cancer.

1.4 APPL proteins

APPL1 was first identified as an AKT2-binding protein in a yeast two-hybrid screen in 1999 [40]. APPL1 was initially called DIP-13α (DCC-interacting protein 13α), as it interacts with the tumor suppressor protein DCC (deleted in colorectal cancer) [41]. APPL proteins, which include APPL1 and APPL2, are named after their unique structure, the multifunctional adaptor proteins that contain a pleckstrin homology (PH) domain, phosphotyrosine binding (PTB) domain, and leucine zipper motif [40]. APPL1 and APPL2 share 54% sequence identity and many identical binding partners. Both are found only in eukaryotes [42]. Briefly, APPL1 consists of the N-terminal Bin1/amphiphysin/rvs167 BAR domain (originally identified as the leucine zipper motif), followed by a pleckstrin homology domain (PH domain), a BPP (region "between PH and PTB domains") domain, a PTB domain, and a C-terminal CC domain [42, 43]. The BAR, PH, and PTB domains are the key functional domains. The BAR and PH domains usually act as a unit involved in sensing and stabilizing membrane curvature and anchor the host proteins to membrane compartments. The PTB domain interacts with phospholipids, receptors such as DCC, and signaling molecules including AKT2. In summary, APPL proteins regulate important physiological processes via their unique domains [44].

APPL1 is a marker of early endosomes that are precursors of classical PI3P-positive endosomes [45]. Depletion of PI3P by PI3K inhibitors leads to the reversion of EEA1-positive endosomes to the APPL1 stage, enlargement of APPL1 endosomes, and enhanced growth factor signaling [45]. APPL proteins are implicated in signaling pathways such as the EGF [46], NF-κB [47], and TGFβ signaling pathways [48]. Through its roles in endocytosis and signal transduction, APPL1 has been reported to mediate proliferation, apoptosis, and migration [44, 49].

2. TGFβ causes Lys63-linked polyubiquitination of TβRI by TRAF6, inducing the formation of the intracellular domain of TβRI (TβRI-ICD), which promotes tumor invasion by inducing the transcription of target genes in the nucleus

We identified the intracellular domain of TβRI by using two different TβRI antibodies: v22, which recognizes the C-terminal part of TβRI; and H100, which was raised against the N-terminal part of TβRI. Upon TGFβ stimulation, the C-terminal fragment of TβRI accumulates in the nucleus. However, the N-terminal part of TβRI still localizes mainly to the cell membrane [50].

We have previously shown that TRAF6 interacts with a consensus binding site in TβRI [22]. Interestingly, TRAF6 is known to cause Lys63-linked polyubiquitination of TβRI, as well as the generation of TβRI-ICD. It has been reported that TNFα-converting enzyme (TACE) induces the cleavage of TβRI through the ERK MAP-kinase pathway [51]. We confirmed that TACE cleaves TβRI by using both an activator of protein kinase C (PKC), which can activate TACE, and an inhibitor of TACE. The TACE cleavage site in TβRI is the Gly-Leu bond at position 120–121, which is close to the transmembrane domain. The G120I mutant has intact kinase activity but does not accumulate in the nucleus in response to TGFβ [50]. PKCζ, which interacts with TRAF6 [52], is required for the formation and nuclear translocation of TβRI-ICD [50].

By immunofluorescence and co-immunoprecipitation, TβRI-ICD has been shown to associate with p300 in the nucleus in a PKCζ-dependent manner. Moreover, p300 mediates the acetylation of TβRI-ICD [50]. In the nucleus,

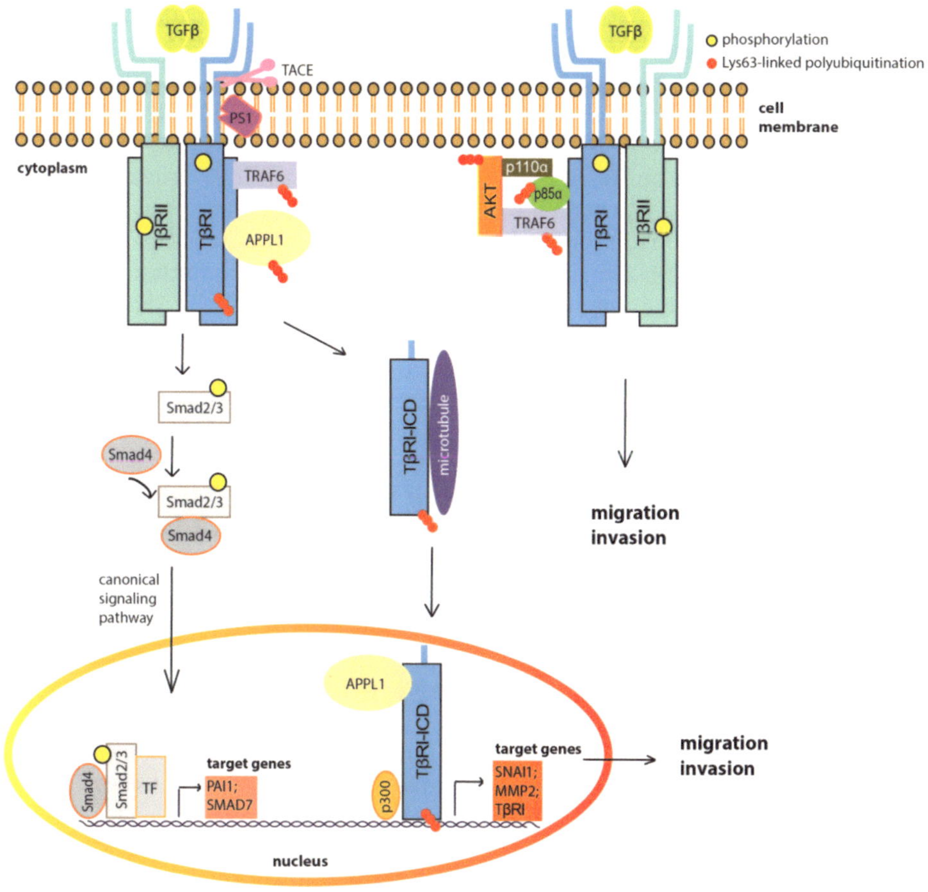

Figure 1.
Proposed model for canonical and TRAF6-mediated non-canonical TGFβ signaling pathways. Upon TGFβ stimulation, constitutively TβRII activates TβRI, leading to the phosphorylation of Smad2 and Smad3. R-Smads, which form a trimeric complex with Smad4, translocate to the nucleus for target genes expression, such as PAI1 and Smad7. In response to TGFβ, TRAF6 induces the formation of TβRI-ICD, which is generated by the proteolytic enzymes TACE and PS1. APPL proteins are necessary for the nuclear translocation of TβRI-ICD. In the nucleus, TβRI-ICD interacts with p300 and promotes tumor invasion indirectly or directly by inducing the transcription of target genes, such as SNAI1, MMP2, and TβRI. TRAF6 also causes the polyubiquitination of p85α, leading to the activation of the PI3K-AKT signaling pathway.

TβRI-ICD regulates the transcription of target genes, such as SNAI1 and MMP2, promoting the invasiveness of cancer cells. Interestingly, the cleavage of TβRI occurs only in malignant prostate cancer cells (PC-3 U), but not in normal primary human prostate epithelial cells. Nuclear accumulation of TβRI-ICD is also observed in prostate cancer, breast cancer, and bladder cancer, suggesting that preventing nuclear translocation of TβRI-ICD could be a new target in cancer treatment [50] (**Figure 1**).

3. TRAF6 induces Lys63-linked polyubiquitination and activation of PS1, leading to the cleavage of TβRI and promoting tumor invasion

Presenilin 1 (PS1) is the catalytic core of the γ-secretase complex, which mediates the cleavage of many cell surface type I transmembrane receptors, such as APP, Notch, and CD44 [53]. TRAF6 is reported to interact with PS1, which enhances the autoubiquitination of TRAF6 [54]. To further investigate the molecular mechanism of TβRI cleavage, we examined the possible involvement of PS1.

TGFβ stimulation enhances the abundance and activity of PS1. PS1 interacts with TβRI in a TRAF6-dependent manner. TRAF6 causes Lys63-linked polyubiquitination of PS1 in response to TGFβ, leading to the activation of PS1. After the initial cleavage of TβRI by TACE, activated PS1 mediates a second cleavage between Val129 and Ile 130 in the transmembrane domain of TβRI, leading to the generation and nuclear translocation of TβRI-ICD [55].

In the nucleus, TβRI-ICD induces its own gene expression to promote cell invasion (**Figure 1**). Experiments using γ-secretase inhibitors showed that PS1 is required for TGFβ-induced cell invasion *in vitro*. Furthermore, γ-secretase inhibitors also reduce the generation of TβRI-ICD and tumor growth in a prostate cancer xenograft model *in vivo*, suggesting a novel therapeutic strategy for cancers [55].

4. Lys178 in TβRI is the acceptor lysine of Lys63-linked polyubiquitination by TRAF6, which is involved in TGFβ-induced invasion and cell cycle regulation

In *in vitro* and *in vivo* ubiquitination assays, TβRI Lys178, the only lysine close to the TRAF6 consensus binding site, has been identified as the acceptor lysine in polyubiquitination by TRAF6. Overexpression of HA-TβRI-K178R inhibits the formation and nuclear translocation of TβRI-ICD in response to TGFβ. The HA-TβRI-K178R mutant has no effect on the kinase activity of TβRI, indicating that it does not interfere with the phosphorylation of Smad2. However, transfection of cells with HA-TβRI-K178R does alter p38 activation [56].

We identified additional genes targeted by nuclear TβRI-ICD by using qRT-PCR. Overexpression of HA-TβRI-K178R changes the expression of genes implicated in invasiveness and cell cycle regulation, such as Vimentin, Twist1, N-cadherin, CCND1, and p73. As expected, the expression of PAI1 is unchanged, due to the intact kinase activity of HA-TβRI-K178R. Fewer cells enter G1 from G0 in HA-TβRI-K178R-transfected cells compared with HA-TβRI-transfected cells after incubation with TGFβ for 48 hours, as CCND1 is poorly regulated in the mutant-transfected cells. PC-3 U cells expressing HA-TβRI-K178R were less invasive than cells expressing HA-TβRI. In summary, the polyubiquitination of TβRI on Lys178 influences both cell cycle regulation and invasion [56].

5. APPL proteins are required for the nuclear translocation of the TGFβ type I receptor intracellular domain

Next, we started to investigate the mechanism of nuclear translocation of TβRI-ICD. As APPL proteins are involved in cargo trafficking from the endosomal membranes to the nucleus after EGF stimulation [46], we considered the possibility that APPL proteins play the same role in the translocation of TβRI-ICD.

The nuclear accumulation of TβRI-ICD in response to TGFβ decreased after APPL1/2 expression was silenced. Moreover, APPL1 overexpression increased the nuclear translocation of TβRI-ICD, indicating that APPL proteins are necessary for the transport of TβRI-ICD into the nucleus. Interestingly, APPL proteins also affect the activation of Smad2 and p38, suggesting that APPL1/2 may play a role in both canonical and non-canonical TGFβ signaling [48].

Using co-immunoprecipitation and an *in vitro* binding assay, we confirmed that APPL1, through its C-terminus, interacts directly with TβRI. TGFβ stimulation enhances the formation of the APPL1-TβRI complex. Moreover, treatment with PI3K inhibitors such as LY294002 and wortmannin enlarges APPL1 early endosomes and prevents the maturation of APPL1 endosomes to EEA1-positive endosomes, and causes increased association of APPL1 with TβRI. In contrast, TβRI kinase activity is not necessary for the interaction between APPL1 and TβRI. Furthermore, endogenous APPL1 has been shown in a nuclear fractionation assay to interact with TβRI-ICD in the nucleus after TGFβ stimulation [48].

It has been reported that APPL1 undergoes Lys63-linked polyubiquitination mediated by TRAF6 in response to insulin in primary mouse hepatocytes [57]. We found that TRAF6 also causes Lys63-linked polyubiquitination of APPL1 after TGFβ stimulation of human prostate (PC-3 U) cells. Of note, TRAF6 is required for both the formation of the APPL1-TβRI complex and the interaction between APPL1 and β-tubulin. In summary, we conclude that APPL proteins are required for the nuclear translocation of TβRI-ICD, possibly via the microtubule system [48] (**Figure 1**).

Nuclear TβRI-ICD promotes the invasion of various cancer cells by inducing the transcription of pro-invasion genes, such as MMP2 and MMP9 [50]. After silencing the expression of APPL1/2, TGFβ-induced invasion is reduced, probably due to a decline in the nuclear accumulation of TβRI-ICD, in both a prostate cancer cell line (PC-3 U) and a breast cancer cell line (MDA-MB-231). MMP2 and MMP9 gene expression also decreases after APPL1/2 knock-down. We also found that APPL1 staining is correlated with a high Gleason Score (indicating the tumor invasiveness and bad prognosis), consistent with previous reports [48, 58]. Interestingly, using an *in situ* proximity ligation assay, we found more APPL1–TβRI-ICD complexes in high-Gleason Score patients. In summary, APPL1–TβRI-ICD is a potential prognostic marker for prostate cancer patients [48].

6. TGFβ activates the PI3K/AKT signaling pathway by TRAF6-mediated polyubiquitination of p85α

It has been reported that TGFβ can activate AKT. However, the detailed mechanism is still unclear. We found that, upon TGFβ stimulation, TβRI forms a complex with AKT and the phosphorylation of AKT correlates with its interaction with TβRI and TRAF6 [59]. As TRAF6 causes Lys63-linked polyubiquitination and activation of AKT upon IGF-1, LPS, and IL-1β stimulation [60], we investigated whether TRAF6 plays the same role in the TGFβ signaling pathway. Using an *in vivo* ubiquitination assay in PC-3 U cells, we demonstrated that TGFβ induces Lys63-linked polyubiquitination of AKT, which is mediated by TRAF6. TGFβ stimulation induces recruitment of the activated-AKT–TRAF6–TβRI complex to cell membrane ruffles. The interaction between TβRI and AKT does not require TβRI kinase activity, but depends on the regulatory subunit of PI3K, p85α. Furthermore, p85α is also involved in the activation and ubiquitination of AKT [59].

The interaction between TRAF6 and p85α is enhanced after TGFβ stimulation. TGFβ induces the Lys63-linked polyubiquitination of p85α in a TRAF6-dependent manner (**Figure 1**). The kinase activities of TβRI and TβRII are not involved in p85α ubiquitination. p85α was found to associate with TβRI upon TGFβ stimulation, but not with TβRII, and TβRI kinase activity is not necessary for the interaction between p85α and TβRI. We found that TGFβ induces PI3K activity in a TRAF6-dependent manner, but independently of TβRI kinase activity, and that TGFβ promotes cell migration and invasion via the PI3K pathway and TRAF6 [59]. Using mass spectrometry and an *in vivo* ubiquitination assay, we identified Lys513 and/or

Lys519 in the iSH2 domain as the major residue(s) of Lys63-linked polyubiquitination of p85α. Overexpression of a K513/K519 double mutant not only suppresses PI3K activity and AKT phosphorylation, but also inhibits cell migration and invasion. Finally, using an *in situ* proximity ligation assay performed in prostate cancer tissue samples, we found that polyubiquitination of p85α is correlated with the aggressiveness of the prostate cancer, suggesting that the polyubiquitination of p85α could be a prognostic marker for this disease [59]. As both the TGFβ and PI3K pathways are deregulated in cancers, finding the link between these two pathways will be important for future cancer research [61].

7. Conclusions

Ubiquitination regulates a broad spectrum of physiological processes, including cell proliferation, apoptosis, differentiation, and others [1, 2]. We have shown that, upon TGFβ stimulation, TRAF6 causes Lys63-linked polyubiquitination of p85α, leading to the activation of the AKT signaling pathway [59]. Moreover, TGFβ, via TRAF6, causes Lys63-linked polyubiquitination of TβRI and its PKCζ-dependent cleavage by TACE [50]. After this initial cleavage by TACE, PS1 is activated by TRAF6-mediated polyubiquitination, which results in a second cleavage of TβRI, by PS1 [55]. APPL proteins are involved in the nuclear translocation of TβRI-ICD [48]. In the nucleus, TβRI-ICD promotes the transcription of pro-invasion genes, such as SNAI1, MMP2, and TβRI itself [50, 55]. TβRI-ICD can be found in cancer cell lines, but not in normal prostate epithelial cell lines or in the normal prostate epithelium [50]. Inhibitors of γ-secretase, which prevent the generation of TβRI-ICD, suppress cell invasion *in vitro* and tumor growth *in vivo*, indicating a possible novel therapeutic target in cancer [55].

Acknowledgements

Fund: CAN 2017/544, Swedish Medical Research Council (2019-01598), Prostatacancerförbundet, King Gustaf V and Queen Victoria's Foundation of Freemasons, Novo Nordic Foundation and Lions Cancer Research Foundation, Umeå University, the County of Västerbotten (RV 933125, RV 73891). The funders did not play a role in manuscript design, data collection, data analysis, data interpretation, or writing of the manuscript.

Conflict of interest

The authors declare no conflict of interest.

Abbreviations

APPL1	adaptor protein phosphotyrosine interaction, PH domain, and leucine zipper containing 1
APPL2	adaptor protein phosphotyrosine interaction, PH domain, and leucine zipper containing 2
EEA1	early endosome antigen 1
ICD	intracellular domain
MAPK	mitogen-activated protein kinase

MAPKKK	mitogen-activated protein kinase kinase kinase
MKK	mitogen-activated protein kinase kinase
MMP	matrix metallopeptidase
NF-κB	nuclear factor kappa B
PC-3 U	prostate cancer-3-Uppsala
PH domain	pleckstrin homology domain
PI3K	phosphatidylinositol-3′-kinase
PKC	protein kinase C
PS1	presenilin 1
SARA	Smad anchor for receptor activation
TACE	TNFα-converting enzyme
TβRI	type I transforming growth factor β receptor
TβRII	type II transforming growth factor β receptor
TGFβ	transforming growth factor β
TNF	tumor necrosis factor alpha
TRAF6	tumor necrosis factor receptor-associated factor 6

Author details

Jie Song and Maréne Landström*
Medical Biosciences, Pathology, Umeå University, Umeå, Sweden

*Address all correspondence to: marene.landstrom@umu.se

References

[1] Mansour MA. Ubiquitination: Friend and foe in cancer. The International Journal of Biochemistry & Cell Biology. 2018;**101**:80-93. DOI: 10.1016/j.biocel.2018.06.001

[2] Chen ZJ, Sun LJ. Nonproteolytic functions of ubiquitin in cell signaling. Molecular Cell. 2009;**33**:275-286. DOI: 10.1016/j.molcel.2009.01.014

[3] Dittmar G, Winklhofer KF. Linear ubiquitin chains: Cellular functions and strategies for detection and quantification. Front Chem. 2020;7:Article 915. DOI: 10.3389/fchem.2019.00915

[4] Huang Q, Zhang X. Emerging roles and research tools of atypical ubiquitination. Proteomics. 2020:201900100. DOI: 10.1002/pmic.201900100

[5] Walczak H, Iwai K, Dikic I. Generation and physiological roles of linear ubiquitin chains. BMC Biology. 2012;**10**:23. DOI: 10.1186/1741-7007-10-23

[6] Iwai K, Fujita H, Sasaki Y. Linear ubiquitin chains: NF-κB signalling, cell death and beyond. Nature Reviews. Molecular Cell Biology. 2014;**15**:503-508. DOI: 10.1038/nrm3836

[7] Spit M, Rieser E, Walczak H. Linear ubiquitination at a glance. Journal of Cell Science. 2019;**132**:jcs208512. DOI: 10.1242/jcs.208512

[8] Swatek KN, Komander D. Ubiquitin modifications. Cell Research. 2016;**26**:399-422. DOI: 10.1038/cr.2016.39

[9] Xie P. TRAF molecules in cell signaling and in human diseases. Journal of Molecular Signaling. 2013;**8**:7. DOI: 10.1186/1750-2187-8-7

[10] Zhu S, Jin J, Gokhale S, Lu AM, Shan H, Feng J, et al. Genetic alterations of TRAF proteins in human cancers. Frontiers in Immunology. 2018;**9**:Article 2111. DOI: 10.3389/fimmu.2018.02111

[11] Ishida T, Mizushima S, Azuma S, Kobayashi N, Tojo T, Suzuki K, et al. Identification of TRAF6, a novel tumor necrosis factor receptor-associated factor protein that mediates signaling from an amino-terminal domain of the CD40 cytoplasmic region. Journal of Biological Chemistry. 1996;**271**:28745-28748. DOI: 10.1074/jbc.271.46.28745

[12] Cao Z, Xiong J, Takeuchi M, Kurama T, Goeddel DV. TRAF6 is a signal transducer for interleukin-1. Nature. 1996;**383**:443-446. DOI: 10.1038/383443a0

[13] Dainichi T, Matsumoto R, Mostafa A, Kabashima K. Immune control by TRAF6-mediated pathways of epithelial cells in the EIME (epithelial immune microenvironment). Frontiers in Immunology. 2019;**10**:Article 1107. DOI: 10.3389/fimmu.2019.01107

[14] Yakymovych I, Yakymovych M, Heldin CH. Intracellular trafficking of transforming growth factor β receptors. Acta Biochimica et Biophysica Sinica. 2018;**50**:3-11. DOI: 10.1093/abbs/gmx119

[15] Batlle E, Massagué J. Transforming growth factor-β signaling in immunity and cancer. Immunity. 2019;**50**:924-940. DOI: 10.1016/j.immuni.2019.03.024

[16] Dijke PT, Arthur HM. Extracellular control of TGFβ signalling in vascular development and disease. Nature Reviews. Molecular Cell Biology. 2007;**8**:857-869. DOI: 10.1038/nrm2262

[17] Moustakas A, Heldin C-H. Mechanisms of TGFβ-induced

epithelial–mesenchymal transition. Journal of Clinical Medicine. 2016;**5**. DOI: 10.3390/jcm5070063

[18] Derynck R, Budi EH. Specificity, versatility, and control of TGF-β family signaling. Science Signalling. 2019;**12**:eaav5183. DOI: 10.1126/scisignal.aav5183

[19] Mu Y, Gudey SK, Landström M. Non-Smad signaling pathways. Cell and Tissue Research. 2012;**347**:11-20. DOI: 10.1007/s00441-011-1201-y

[20] Heldin C-H, Landström M, Moustakas A. Mechanism of TGF-β signaling to growth arrest, apoptosis, and epithelial-mesenchymal transition. Current Opinion in Cell Biology. 2009;**21**:166-176. DOI: 10.1016/j.ceb.2009.01.021

[21] Heldin C, Moustakas A. Signaling receptors for TGF-β family members. Cold Spring Harbor Perspectives in Biology. 2016;**8**. DOI: 10.1101/cshperspect.a022053

[22] Sorrentino A, Thakur N, Grimsby S, Marcusson A, von Bulow V, Schuster N, et al. The type I TGF-β receptor engages TRAF6 to activate TAK1 in a receptor kinase-independent manner. Nature Cell Biology. 2008;**10**:1199-1207. DOI: 10.1038/ncb1780

[23] Yamashita M, Fatyol K, Jin C, Wang X, Liu Z, Zhang YE. TRAF6 mediates Smad-independent activation of JNK and p38 by TGF-β. Molecular Cell. 2008;**31**:918-924. DOI: 10.1016/j.molcel.2008.09.002

[24] Doherty GJ, McMahon HT. Mechanisms of endocytosis. Annual Review of Biochemistry. 2009;**78**:857-902. DOI: 10.1146/annurev.biochem.78.081307.110540

[25] Chen Y-G. Endocytic regulation of TGF-beta signaling. Cell Research. 2009;**19**:58-70. DOI: 10.1038/cr.2008.315

[26] Balogh P, Katz S, Kiss AL. The role of endocytic pathways in TGF-β signaling. Pathology Oncology Research. 2013;**19**:141-148. DOI: 10.1007/s12253-012-9595-8

[27] Hu CT, Wu JR, Wu WS. The role of endosomal signaling triggered by metastatic growth factors in tumor progression. Cellular Signalling. 2013;**25**:1539-1545. DOI: 10.1016/j.cellsig.2013.03.022

[28] Yu Y, Feng XH. TGF-β signaling in cell fate control and cancer. Current Opinion in Cell Biology. 2019;**61**:56-63. DOI: 10.1016/j.ceb.2019.07.007

[29] David CJ, Massagué J. Contextual determinants of TGFβ action in development, immunity and cancer. Nature Reviews. Molecular Cell Biology. 2018;**19**:419-435. DOI: 10.1038/s41580-018-0007-0

[30] Thorpe LM, Yuzugullu H, Zhao JJ. PI3K in cancer: Divergent roles of isoforms, modes of activation and therapeutic targeting. Nature Reviews. Cancer. 2015;**15**:7-24. DOI: 10.1038/nrc3860

[31] Hoxhaj G, Manning BD. The PI3K–AKT network at the interface of oncogenic signalling and cancer metabolism. Nature Reviews. Cancer. 2020;**20**:74-88. DOI: 10.1038/s41568-019-0216-7

[32] Madsen RR, Vanhaesebroeck B. Cracking the context-specific PI3K signaling code. Science Signalling. 2020;**13**:eaay2940. DOI: 10.1126/scisignal.aay2940

[33] Vallejo-Díaz J, Chagoyen M, Olazabal-Morán M, González-García A, Carrera AC. The opposing roles of PIK3R1/p85α and PIK3R2/p85β in Cancer. Trends in Cancer. 2019;**5**:233-244. DOI: 10.1016/j.trecan.2019.02.009

[34] Manning BD, Toker A. AKT/PKB signaling: Navigating the network. Cell. 2017;**169**:381-405. DOI: 10.1016/j.cell.2017.04.001

[35] Chan C-H, Jo U, Kohrman A, Rezaeian AH, Chou P-C, Logothetis C, et al. Posttranslational regulation of Akt in human cancer. Cell & Bioscience. 2014;**4**:59. DOI: 10.1186/2045-3701-4-59

[36] Bakin AV, Tomlinson AK, Bhowmick NA, Moses HL, Arteaga CL. Phosphatidylinositol 3-kinase function is required for transforming growth factor β-mediated epithelial to mesenchymal transition and cell migration. The Journal of Biological Chemistry. 2000;**275**:36803-36810. DOI: 10.1074/jbc.M005912200

[37] Lamouille S, Derynck R. Cell size and invasion in TGF-β-induced epithelial to mesenchymal transition is regulated by activation of the mTOR pathway. The Journal of Cell Biology. 2007;**178**:437-451. DOI: 10.1083/jcb.200611146

[38] Zhang L, Zhou F, Dijke PT. Signaling interplay between transforming growth factor-β-receptor and PI3K/AKT pathways in cancer. Trends in Biochemical Sciences. 2013;**38**:612-620. DOI: 10.1016/j.tibs.2013.10.001

[39] Yi JY, Shin I, Arteaga CL. Type I transforming growth factor β receptor binds to and activates phosphatidylinositol 3-kinase. The Journal of Biological Chemistry. 2005;**280**:10870-10876. DOI: 10.1074/jbc.M413223200

[40] Mitsuuchi Y, Johnson SW, Sonoda G, Tanno S, Golemis EA, Testa JR. Identification of a chromosome 3p14.3-21.1 gene, APPL, encoding an adaptor molecule that interacts with the oncoprotein-serine/threonine kinase AKT2. Oncogene. 1999;**18**:4891-4898. DOI: 10.1038/sj.onc.1203080

[41] Liu J, Yao F, Wu R, Morgan M, Thorburn A, Finley RL, et al. Mediation of the DCC apoptotic signal by DIP13α. The Journal of Biological Chemistry. 2002;**277**:26281-26285. DOI: 10.1074/jbc.M204679200

[42] Liu Z, Xiao T, Peng X, Li G, Hu F. APPLs: More than just adiponectin receptor binding proteins. Cellular Signalling. 2017;**32**:76-84. DOI: 10.1016/j.cellsig.2017.01.018

[43] Li J, Mao X, Dong LQ, Liu F, Tong L. Crystal structures of the BAR-PH and PTB domains of human APPL1. Structure. 2007;**15**:525-533. DOI: 10.1016/j.str.2007.03.011

[44] Deepa SS, Dong LQ. APPL1: Role in adiponectin signaling and beyond. American Journal of Physiology. Endocrinology and Metabolism. 2009;**296**:E22-E36. DOI: 10.1152/ajpendo.90731.2008

[45] Zoncu R, Perera RM, Balkin DM, Pirruccello M, Toomre D, De Camilli P. A phosphoinositide switch controls the maturation and signaling properties of APPL endosomes. Cell. 2009;**136**:1110-1121. DOI: 10.1016/j.cell.2009.01.032

[46] Miaczynska M, Christoforidis S, Giner A, Shevchenko A, Uttenweiler-Joseph S, Habermann B, et al. APPL proteins link Rab5 to nuclear signal transduction via an endosomal compartment. Cell. 2004;**116**:445-456. DOI: 10.1016/s0092-8674(04)00117-5

[47] Hupalowska A, Pyrzynska B, Miaczynska M. APPL1 regulates basal NF-κB activity by stabilizing NIK. Journal of Cell Science. 2012;**125**:4090-4102. DOI: 10.1242/jcs.105171

[48] Song J, Mu Y, Li C, Bergh A, Miaczynska M, Heldin C-H, et al. APPL proteins promote TGFβ-induced nuclear transport of the TGFβ type I receptor intracellular domain. Oncotarget.

2016;7:279-292. DOI: 10.18632/oncotarget.6346

[49] Diggins NL, Webb JD. APPL1 is a multi-functional endosomal signaling adaptor protein. Biochemical Society Transactions. 2017;**45**:771-779. DOI: 10.1042/BST20160191

[50] Mu Y, Sundar R, Thakur N, Ekman M, Gudey SK, Yakymovych M, et al. TRAF6 ubiquitinates TGFβ type I receptor to promote its cleavage and nuclear translocation in cancer. Nature Communications. 2011;**2**:330. DOI: 10.1038/ncomms1332

[51] Liu C, Xu P, Lamouille S, Xu J, Derynck R. TACE-mediated ectodomain shedding of the type I TGF-β receptor downregulates TGF-β signaling. Molecular Cell. 2009;**35**:26-36. DOI: 10.1016/j.molcel.2009.06.018

[52] Feng Y, Longmore GD. The LIM protein Ajuba influences Interleukin-1-induced NF-κB activation by affecting the assembly and activity of the protein kinase Cζ/p62/TRAF6 signaling complex. Molecular and Cellular Biology. 2005;**25**:4010-4022. DOI: 10.1128/mcb.25.10.4010-4022.2005

[53] McCarthy JV, Twomey C, Wujek P. Presenilin-dependent regulated intramembrane proteolysis and γ-secretase activity. Cellular and Molecular Life Sciences. 2009;**66**:1534-1555. DOI: 10.1007/s00018-009-8435-9

[54] Powell JC, Twomey C, Jain R, McCarthy JV. Association between Presenilin-1 and TRAF6 modulates regulated intramembrane proteolysis of the p75NTR neurotrophin receptor. Journal of Neurochemistry. 2009;**108**:216-230. DOI: 10.1111/j.1471-4159.2008.05763.x

[55] Gudey SK, Sundar R, Mu Y, Wallenius A, Zang G, Bergh A, et al. TRAF6 stimulates the tumor-promoting effects of TGFβ type I receptor through polyubiquitination and activation of presenilin. Science Signalling. 2014;**7**:ra2. DOI: 10.1126/scisignal.2004207

[56] Sundar R, Gudey SK, Heldin C-H, Landström M. TRAF6 promotes TGFβ-induced invasion and cell-cycle regulation via Lys63-linked polyubiquitination of Lys178 in TGFβ type I receptor. Cell Cycle. 2015;**14**:554-565. DOI: 10.4161/15384101.2014.990302

[57] Cheng KKY, Lam KSL, Wang Y, Wu D, Zhang M, Wang B, et al. TRAF6-mediated ubiquitination of APPL1 enhances hepatic actions of insulin by promoting the membrane translocation of Akt. The Biochemical Journal. 2013;**455**:207-216. DOI: 10.1042/BJ20130760

[58] Johnson IRD, Parkinson-Lawrence EJ, Keegan H, Spillane CD, Barry-O'Crowley J, Watson WR, et al. Endosomal gene expression: A new indicator for prostate cancer patient prognosis? Oncotarget. 2015;**6**:37919-37929. DOI: 10.18632/oncotarget.6114

[59] Hamidi A, Song J, Thakur N, Itoh S, Marcusson A, Bergh A, et al. TGF-β promotes PI3K-AKT signaling and prostate cancer cell migration through the TRAF6-mediated ubiquitylation of p85α. Science Signalling. 2017;**10**:eaal4186. DOI: 10.1126/scisignal.aal4186

[60] Yang W, Wang J, Chan C, Lee S, Campos DA, Lamothe B, et al. The E3 ligase TRAF6 regulates Akt ubiquitination and activation. Science. 2009;**325**:1134-1138. DOI: 10.1126/science.1175065

[61] Song J, Landström M. TGFβ activates PI3K-AKT signaling via TRAF6. Oncotarget. 2017;**8**:99205-99206. DOI: 10.18632/oncotarget.22275

Chapter 4

Ubiquitination and Deubiquitination in Melanoma Research and Clinically Relevant Outcomes

Jia Guo and Jianglin Guo

Abstract

Malignant melanoma is one of the most invasive tumors with increasing mortality, low overall survival rates and limited effective therapeutic strategies. Ubiquitination is a post-translational protein modification, which is regulated by a series of ubiquitination-associated enzymes. Ubiquitination plays a critical role in diverse pathophysiological activities of cellular and participates in the pathogenesis of various cancers, including melanoma. This study aims to provide a conclusive of ubiquitination and deubiquitination, and their potential clinical application value in melanoma in the following aspects: melanoma pathogenesis-related components and processes in the ubuiquitin-proteasome system (UPS), ubiquitination in melanoma immunological microenvironment modulation, ubiquitination of key transcription factors in melanoma and melanoma therapeutic strategy via targeting the UPS.

Keywords: ubiquitination, deubiquitinating enzymes, melanoma, pathogenesis, application

1. Introduction

Malignant melanoma is one of the most invasive tumors with increasing mortality, low overall survival rates, and limited effective therapeutic strategies [1]. Although melanoma is the third most prevalent skin cancer, the two other skin malignancies, basal cell and squamous cells, are the malignant [2]. A variety of factors, including genetic mutations, sun exposure, and poor lifestyle habits, are involved in the development of melanoma [3].

There is a dynamic protein balance in cells to maintain homeostasis for cell and organism. Intracellular protein degradation by two pathways, autophagy and lysosomal degradation pathway and the ubiquitin-proteasome pathway, is primarily involved in tumor growth. Ubiquitination is one of post-translational modifications of most vital proteins. Ubiquitin, a closely conserved small protein composed of 76 amino acids, is link with the ubiquitin-activating enzyme (E1), the ubiquitin-conjugating enzyme (E2), and the ubiquitin ligase (E3) [4]. Specifically, mono-ubiquitination was considered as only one single ubiquitin bond to the lysine, while poly-ubiquitination was considered as ubiquitin chains attached to the lysine [5]. Then the ubiquitinated proteins are transported to the 26S proteasome for degradation. Ubiquitination is involved in the development of different tumors by regulation important genes or signaling pathways. However, the ubiquitination process can be reversed by the deubiquitinating enzymes (DUBs) via cleaving ubiquitin chains from substrates to prevent protein degradation, which participates in a wide range of cellular signaling pathways, such as the apoptosis, cell cycle, autophagy, DNA damage, inflammation signaling, and protein downregulation [6]. Up to date, there were reported over 600 E3 ubiquitin ligase and 100 DUBs [7].

A significant number of studies have confirmed that ubiquitination and de-ubiquitination play a critical role in melanoma pathogenesis, and have indicated that the key molecular goal of the mechanism may be the therapeutic strategies for the treatment of melanoma. Here, we provide a conclusive introduction about protein ubiquitin modification in relative genes, signaling pathways, and in immune system in melanoma pathogenesis, which concludes the latest DUB studies in melanoma. Besides, we summarize potential therapeutic targets of ubiquitination and de-ubiquitination in melanoma.

2. Melanoma pathogenesis related components and processes of the UPS

2.1 Fbxw7

The F-box/WD repeat-containing protein 7 (Fbxw7) belongs to the F-box protein family, which is the component of an SCF E3 ubiquitin ligase [8]. Fbxw7 is considered to be a tumor suppressor gene [9]. The degradation of Fbxw7 results in accumulation of its substrates, leading to oncogenesis. In a study, the mutation prevalence was found to be 8.1% FBXW7 in melanoma through exome sequencing in a cohort of 103 melanomas. A potential therapeutic approach for melanoma could be the loss and mutation of FBXW7 in melanoma contributing to prolonged activation of NOTCH and targeting NOTCH signaling [10]. FBXW7 deficiency can also unleash heat shock factor 1 (HSF1) and then result in melanoma invasion and metastasis [11]. Meanwhile, FBXW7 can regulate the melanoma metastasis through activating the MAPK/ERK signaling [12]. The microphthalmia-associated transcription factor (MITF) is a key regulator of melanocyte development, differentiation, and melanoma biology [13, 14]. FBXW7 is recognized as a regulator of MITF via post-transcriptional mechanisms [15].

2.2 SKP2

S-phase kinase-associated protein 2 (Skp2), also be called FBXL1 or p45, is also a member of the F-box proteins [16]. Skp2 is characterized as a cancer-related protein. In general, in primary melanoma and metastasis melanoma, SKP2 is significantly up-regulated, which is related to the prognosis, as it is reported that nuclear Skp2 expression is strongly associated with a lower survival rate during melanoma [17, 18]. In tumorigenesis, Skp2 stabilizes the MTH1 expression via K63-linked polyubiquitination, and then promotes melanoma cell survival by protecting DNA integrity upon pharmacologic oxidative stress [19]. Meanwhile, skp2 has a direct interaction with melanoma antigen-A11 (MAGE-A11), which may boost Skp2-mediated degradation of cyclin A [20].

2.3 HACE1

HECT domain and ankyrin repeat-containing E3 ubiquitin protein ligase 1 (HACE1), has been showed to act as a tumour suppressor gene in various kinds of cancers [21]. There is a significantly downregulation of HACE1 in colorectal cancer (CRC), and the decreased expression is highly associated with poor clinical features of patients. HACE1 inhibits YAY1 signaling and then can reverse EMT in CRC [22]. Loss of HACE1 activates RAC-family GTPases to mediate oxidative stress that increases genotoxic cellular ROS generation and then results in lung tumor formation [23]. Even though HACE1 behaves as an anti-oncogene in most reports, its function in melanoma may be cell-specific tumorigenesis. HACE1 plays a pro-oncogenic role in melanoma by regulating fibronectin (FN) secretion and K27 ubiquitination of FN [24].

2.4 ITCH

ITCH, a member of HECT-type ubiquitin E3 ligases, plays a significantly role in regulating cell growth and apoptosis [25]. In melanoma, ITCH mediates BRAF polyubiquitination through the K27-linkage result in sustained activation of BRAF/MEK/ERK signaling, which leads to the survival of melanoma cells [26]. Moreover, ITCH can be regulated by microRNAs (miRNAs), such as miR-10b and miR-520f, and then be involved in the melanoma proliferation and metastasis [27, 28].

2.5 UBE2C

UBE2C belonging to the E2 family is operating in combination with the anaphase-promoting complex/cyclosome (APC/C) E3 ligase. It regulates the cell cycle through mitosis via destructing mitotic cyclin B1 [29]. Silence of UBE2C induces G2/M phase arrest of melanoma cells by suppressing both the level and the activity of M-phase-promoting factor (MPF), a complex consisting of CDK1 and cyclin B1 [30, 31].

2.6 UBE2S

Ubiquitin-conjugating enzyme E2S (UBE2S) belongs to the E2 protein family, and is involved in development of various cancers. Recently, it has been shown that UBE2S plays a vital role in regulating DNA damage-induced transcriptional silencing, by catalyzing Lys11-linkage ubiquitination [32]. Another recent research showed that UBE2S is overexpressed in melanoma, and the expression was significantly related to the cancer staging and grading, with a higher magnitude found for tumor node metastasis staging T4. Moreover, silence of UBE2S may cause melanoma cell proliferation inhibition via inducing cell cycle G1/S phase arrest, and cell apoptosis. In BALB/C nude mice, shUBE2S can suppress tumor growth and inhibit epithelial-mesenchymal transition (EMT) [33].

2.7 MKRN2

Makorin ring finger protein 2 (MKRN2) is known as a novel ubiquitin E3 ligase, and is capable of targeting the p65 subunit of NF-κB [34]. Research indicates that there is a greater expression of MKRN2 in melanoma cell lines relative to normal skin cell lines. The silence of MKRN2 can inhibit melanoma cell growth in a P53-dependent manner. Moreover, MKRN2 can interact with ubiquitylated P53 [35]. This study suggests that MKRN2 may be a potential therapeutic target for melanoma.

2.8 Ub-like proteins

There are also several ubiquitin-like proteins (UBLs) in addition to Ub, such as NEDD8 (neural precursor cell expressed, developmentally down-regulated8), SUMO (small ubiquitin-like modifiers), and ISG15 (interferon-stimulated gene 15).

NEDD8 mediates the stabilization of various proteins, and plays a significant role in the incidence and development of malignant melanoma. NEDD8 is a ubiquitin-like protein composed of 81 amino acids, with around 60% of the sequence that is the same as ubiquitin [36]. The covalent binding of NEDD8 to substrates is known as neddylation. Similar to the ubiquitination, an enzyme cascade is needed for this progression. Neddylation is involved in protein ubiquitination, and is closely associated with the degradation of certain proteins in the cell cycle and apoptosis-related factors [37]. Cullin is one of the most researched neddylation substrates [38]. Besides, studies have also investigated that NEDD8 substrates are diverse. Some proteins can be modified by NEDD8, including p53 [39], MDM2 [40], and VHL [41]. UBA3, as the subunit of NEDD8-activating enzyme, plays a critical role in the linkage of NEDD8 with cullin proteins. Previous studies have shown that in highly proliferative cell lines, NEDD8 conjugation is up-regulated and increased in melanoma cell lines [42]. After knockdown of UBA3, the proliferation of M14 melanoma cells was suppressed both in vitro and in vivo. Hence, interference of the neddylation might offer a hopeful method for melanoma therapy [43].

SUMO has been described to alter protein interactions rather than directly involving in protein degradation [44]. Sumoylation involves a 3-step pathway analogous to the ubiquitination pathway. Dysregulation of sumoylation has been implicated in multiple cancers, including melanomas. Ubc9, the single SUMO E2 conjugating enzyme, is overexpressed in advanced-stage melanomas where it protects melanoma cells from chemotherapy-induced apoptosis [45]. Moreover, SUMOylation-defective MITF germline mutation may be more susceptible to melanoma [46].

ISG15, a ubiquitin-like modifier, is implicated in both tumor oncogenic and suppressive programs [47]. It is activated by a three steps enzymatic cascade consisting of a specific E1-activating enzyme (UBE1L), E2 conjugating enzyme (typically UBCH8) and E3 ligase (commonly HERC5A), which promotes ISG15 transfer to protein substrates [48]. Previous study shows that ISG15 can be removed from its target proteins by USP18 and then the effects of ISGylation was reversed [49, 50]. A study identifies PTEN as a new substrate of the ISGylation post-translational modification pathway and USP18 can regulate PTEN stability. Inhibition of ISGylation may be a therapeutically relevant in melanoma [47].

2.9 Deubiquitinating enzymes (DUBs)

To date, several DUBs have confirmed to be consistent with melanoma tumorigenesis and metastasis. USP54 is overexpressed in intestinal stem cells, and is defined to promote cancer progression and regulate embryonic development and normal growth of adult mice. USP54 upregulates in melanoma, the loss of USP54 is dispensable for metastasis of melanoma cells [51]. An IFN stimulated to regulate type-I IFN signaling in the anti-viral immune response has been reported to be USP18 [52]. It is also reported that IFN-γ can stimulate USP18 protein expression in melanoma cells. Through IFN-γ-induced USP18 expression in melanoma cells and -regulated CTL CD8 + immune cell activity in the tumor microenvironment, endogenous IFN-γ signaling influences melanoma tumorigenesis [53]. In 2014, Harish Potu et al. reported that USP5 mediates the change in ubiquitinylated protein content and unanchors Ub chains in BRAF mutant cells treated with vemurafenib. BRAF can activate USP5, contributing by suppressing p53 and FAS induction, to inhibit cell cycle checkpoint regulation and apoptosis [54]. In 2018, USP4 upregulation in melanoma, especially in metastatic melanoma, was discovered by Weinan Guo et al. The archive of TCGA skin cutaneous melanoma (SKCM) confirms this finding. USP4 can protect melanoma cells from cisplatin-induced apoptosis in a p53-dependent manner. Moreover, USP4 up-regulation plays an important role in melanoma invasion and migration by promoting EMT [55]. The USP15 knockdown lowers the expression of MDM2 in melanoma cells, and then leads to upregulation of p53 and MDM2 target genes p21 and Puma. Moreover, Usp15−/− melanoma mice models have an increased frequency of CD8+ effector T cells tumor-infiltrating [56].

Ubiquitin specific peptidase 9, X-linked (Usp9x), a member of the USP family, is upregulated in many cancers, which has a positive and negative impact on tumorigenicity depending on the various forms of cancer [57–59]. A study shows that the growth of melanoma cells can be inhibited by Usp9x loss. The Ets-1 proteasomal, abased site-specific de-ubiquitination, is inhibited by Usp9x, which leads to Ets-1 aggregation and increases tumorigenicity of melanoma [60]. Moreover, in malignant melanoma, about 15–20% of NRAS mutations have been identified [61]. Harish Potu et al. also revealed that inhibition of BRAF and/or MEK kinase pathway can increase Ets-1 expression. The increased Ets-1 expression upregulates NRAS levels by activating the NRAS promoter. In all, Usp9x plays a critical role in Ets-1 regulation and melanoma tumorigenicity through mediating NRAS transcription [60]. UCHL1 (ubiquitin C-terminal hydrolase 1) belongs to the ubiquitin carboxy terminal hydrolase family of DUBs. It catalyzes hydrolysis of C-terminal ubiquitin esters to regulate protein degradation [62]. Eun Young Seo et al. have investigated that UCHL1 influences melanogenesis by regulating stability of MITF in human melanocytes, which provides a framework for the further researches to evaluate potent therapeutic approaches for melanoma and other dyspigmentation disorders [63]. BAP1 (BRCA1-associated protein-1) belongs to the UCH subfamily of DUBs, and is known as a tumor suppressor gene [64, 65]. BAP1 mutations were first identified in a small number of lung and breast cancer samples, and have recently been described as leading to the pathogenesis of melanoma [66, 67]. The germline mutations in BAP1 are more prone to malignant melanoma [68]. In 2010, a study reported that 84% of inactivating somatic BAP1 mutations were identified in metastasizing uveal melanomas, including 15 premature protein termination mutations, and six affecting their ubiquitin UCH domains, which were associated with a decrease in BAP1 mRNA level [69]. However, in cutaneous melanoma, the germline mutations in BAP1 were less than 1% and its effect was unknown [70]. A recent study reported that low BAP1 mRNA predicted a better OS in older than 50 years cutaneous melanoma patients after adjusting for ulceration or Breslow depth [71]. The different function of BAP1 in cutaneous melanoma and uveal melanoma needs to be studied further.

3. Ubiquitination in melanoma immunological microenvironment modulation

Tumor microenvironment (TME) is consisted of cancer cells, cancer-associated fibroblasts, immune cells, and stromal cells. TME emerges as a key mechanism that mediate tumor progression [72]. A previous study reported that protein ubiquitylation plays a critical role in modulating immune responses and TME [73].

The Cbl proteins are a family of ubiquitin ligases (E3s). Cbl-b, a member of the family, functions as a negative regulator that regulates CD8 T cells costimulatory pathway and natural killer cell function [74]. In recent years, Cbl-b prone to be one of the hotspot targets of tumor immunotherapy because Cbl-b deletion can cause spontaneous or induced autoimmune call, and Cbl-b overexpression can result in the tumor immune tolerance in infiltrated lymphocytes in TME [75]. A study shows that NK cells knocking down of Cbl-b, or targeting its E3 catalytic activity, inhibit the progression of melanomas and distant melanoma metastases. Moreover, compared with WT T cells, Cbl-b−/− CD8+ and CD4+ T cell proliferation are highly suppressed by a recombinant PD-L1 Ig, and IFN-γ production is significantly less suppressed. Cbl-b deficiency in mice seems to cause a functional resistance of NK cells and T cells to PD-L1/PD-1-mediated immune suppression [76]. Adoptive cell therapy (ACT) with autologous T cells can enforce the immune-mediated tumor cell killing, and show a promising result in various types of cancer treatments [77, 78]. However, the therapeutic efficacy of ACT is still limited because of the tumor-bearing host immune-evasion mechanisms, such as the secretion of transforming growth factor beta (TGFβ) or accumulation of Treg cells, both of which severely dampen the activation, expansion, and tumor homing of CD8+ T cells [79]. Another study reveals that silencing cbl-b reduces TGFβ sensitivity *in vitro* and enhances anti-tumor effects *in vivo*. Adoptive transfer of Cbl-b-silenced CD8+ T lymphocytes augments tumor vaccine to suppress tumor growth and prolong the survival in a B16F10 melanoma model [80].

FBXO38 belongs to the SCF family of E3 ubiquitin ligase of PD-1, and mediates Lys48-linked poly-ubiquitination and substrate proteasome degradation [81]. Previous research investigates that FBXO38 mediates PD-1 ubiquitination and maintains the anti-tumour activity of T cells in melanoma cells [82]. It offers an alternative method to block the PD-1 and highlights the clinical potential of the regulation of anti-tumour immunity through ubiquitination of FBXO38.

SIAH2, potent E3 RING finger ubiquitin ligases, mediates the cell cycle, apoptosis, and DNA repair regulation through targeting subsequent related proteins [83]. Previous study finds that hypoxia activates Siah2 E3 ligase, and then enhances the Warburg effect and pro-tumor immune response via degrading nuclear respiratory factor 1 (NRF1) through ubiquitination on lysine 230 [84]. A recent study reveals the effect and mechanism of Siah2 on the T cells and immune therapy. As is shown in this article, in the one way, Siah2-deficient mice suppress melanoma growth, increase the infiltration of T effector cells, and decrease number of FOXP3+ Treg cells. Inhibition of Siah2−/− melanoma cell proliferation is p27 dependent. Moreover, Siah2−/− BM-transplanted mice inhibit the melanoma growth, which may be a clinical potential of new adoptive cell therapy. On the other hand, loss of Siah2 exhibits synergy with anti-PD1 therapy in melanoma.

In addition to Ub, lots of Ub-related proteins display an immune regulation function in melanoma. A family of Toll-like receptors (TLRs) involves in the recognition of microbial components and regulates innate immune responses [85, 86]. TNFAIP3 (TNF-α induced protein 3), an ubiquitin-editing enzyme, can negatively regulate the TLRs via function as an ubiquitin-editing molecule [87]. E3 ligase NEDD4 mediates the function of immune regulation. Silence of NEDD4 inhibits FOXP3+ Treg cells through mediating GITR degradation, and then contributes to melanoma progression [88]. A previous study finds that USP15 was highly expressed in immune cells through analysis of the BioGPS database. In naïve CD4+ T cells, loss of USP15 stimulates the TCR + CD28 to produce cytokines, such as interleukin 2 (IL-2) and interferon-γ (IFN-γ). Moreover, USP15 inhibits the naïve CD4+ T cell activation and suppresses TH1 differentiation. MDM2, which is recognized as substrate protein of USP15, targets a T cell transcription factor, NFATc2, and negatively regulates T cell activation, which was independent of p53. Later, the author testifies the function of USP15 in B16F10 melanoma models. This study investigated that USP15−/− mice increase IFN-γ + CD4+ T cell infiltration to the tumors, and deficiency of USP15 reduces melanoma tumors size and tumor-induced lethality [89].

4. Ubiquitination of key transcription factors in melanoma

4.1 Ubiquitination of p53

The tumor suppressor protein p53 is a transcription factor that can affect cell proliferation by regulating the expression of its target protein [90]. P53 interacts with E3 ligase MDM2 in the nucleus, and is transferred from the nucleus to the cytoplasm following ubiquitin, resulting in proteasome degrading [91]. In 2003, Leonard Girnita et al. discovered that inhibition of p53 leads to ubiquitination and down-regulation of the IGF-1R in human malignant melanoma cells. This impact was independent of the p53 status (wild type or mutated) but can be rescued by coinhibition of MDM2. Mdm2 serves as a ligase in ubiquitination of the IGF-1R [92]. Unlike other solid tumors, malignant melanomas retain the expression of wild-type p53 and typically lack p53 mutations [93, 94]. Adil Anwar et al. reveal that the wild-type p53 is the target for the ubiquitin-proteasomal pathway (UPP) degradation. The residues serine 15 and serine 20 are also essential for the binding of MDM2, which control p53 destruction via UPP pathway. In this article, p53 stabilization mediated by UPP inhibitors is independent of phosphorylation at residues serine 15 and serine 20 of p53 in melanoma cells [95]. MKRN2 is recognized as a novel ubiquitin E3 ligase targeting the p65 subunit of NF-κB to negatively regulate inflammatory responses [96]. A recent study indicates that the MKRN2 expression increases in the human melanoma cell lines, and silence of this gene leads to the suppression of melanoma proliferation by upregulation of p53. To investigate the mechanism of this effect, authors take co-immunoprecipitation and glutathione S-transferase pulldown assays to confirm the interaction of MKRN2 with p53 and take *in vitro* ubiquitination assays to study the ubiquitination of p53 by MKRN2. The result shows that MKRN2 interacts with p53, and ubiquitylates p53, leading to the influence of melanoma cell proliferation [97].

4.2 Ubiquitination of c-Myc

The transcription factor c-Myc plays an important role in cell proliferation and differentiation, cell cycle, metabolism, and apoptosis [98]. C-Myc is a protein that is very unstable and vulnerable to degradation in a proteasome-dependent manner. Research has identified the E3 ligase of c-Myc in melanoma. Also, c-Myc can be specifically bound by the E3 ligase SKP2 [99].

5. Melanoma therapeutic strategy via targeting the ubiquitin-proteasome system (UPS)

In protein degradation and melanoma pathogenesis, the UPS plays a crucial role, as shown above. The pathogenesis of malignant melanoma leads to genetic changes, irregular expression, or dysfunction [100]. Hence, targeting the UPS may be a potential therapeutic strategy for melanoma. Currently, many small molecule inhibitors targeting different components of the UPS, including the proteasome, E3 ligases, E1 enzymes, E2 enzymes, ubiquitin-like proteins, and DUBs, have been developed [101].

Bortezomib is the first proteasome inhibitor approved by FDA, which was originally used for multiple myeloma treatment [102]. However, due to the clinical safety, the study in other cancer researches, including melanoma, has been discontinued. Compared to the proteasome inhibitor bortezomib, drugs targeting a particular E3 ubiquitin ligase are expected to have better selectivity with less associated toxicity relative to the proteasome inhibitor bortezomib [103]. MDM2 is an E3 ubiquitin ligase with the ability to regulate tumor suppressor p53 and potentiate Notch signaling by degrading Numb [104, 105]. Nutlin-3a, an imidazoline compound, has been generally known as a MDM2 inhibitor. Nutlin-3a can suppress melanoma and other cancers, including retinoblastoma, leukemia, and neuroblastoma [106]. Meanwhile, WIP1 inhibitor (WIP1i), GSK2830371, can enhance p53-mediated tumor suppression by MDM2–p53 inhibitors, nutlin-3, RG7388, and HDM201 in cutaneous melanoma [107]. Therefore, more findings from the phase I clinical trials are needed to evaluate whether there exist any significant side effects.

Besides, Siah2 is known as a RING finger E3 ubiquitin ligase. Inhibition of Siah2 activity using a peptide is reported to be able to weaken its effect on hypoxia, effectively leading to melanoma metastasis inhibition, while suppression of Siah2 activities prevents the tumorigenicity of melanoma by disrupting Ras/MAPK signaling pathways [108]. Menadione (MEN), also known as vitamin K3, is a quinone used for cancer chemotherapeutic agents. A recent research identifies MEN as a novel Siah2 inhibitor, which attenuates hypoxia and MAPK signaling, and blocks melanoma tumorigenesis [109]. This study revealed that targeting Siah2 by MEN may be a new therapeutic strategy in melanoma treatment.

Cullin-RING ligases (CRLs) are a subgroup of the E3 ligases, and play an important role in the degradation of oncology relative proteins. It is activated by the neddylation pathway, such as NEDD8 conjugation [110]. The NEDD8-activating enzyme (NAE) is a critical regulator of the neddylation pathway [111]. Pevonedistat (TAK-924/MLN4924) is reported as a first-in-class, small-molecule inhibitor of NAE. Previous preclinical studies reveal that Pevonedistat is associated with tumor growth inhibition of a range of cell lines and primary human cancer cells derived from solid tumors, including malignant melanoma [112, 113]. A phase I study of Pevonedistat about patients with advanced solid tumors was undertaken. The results find that, in nine melanoma patients, one achieved a partial response (PR) while another 8 patients achieved stable disease (SD) lasting 6 months [114]. In addition, another phase I study (NCT01011530) was conducted to assess the safety, pharmacokinetics (PK), pharmacodynamic (PD), and antitumor activity of Pevonedistat in metastatic melanoma patients. The maximum tolerated dose (MTD) is reported as 209 mg/m^2. Most patients have a well toleration to Pevonedistat, only 16% patients experience drug-related serious adverse event (SAE), such as drug-related grade 4 acute renal failure, grade 3 myocarditis, and grade 3 small intestinal obstruction. At the end of the research, the research results show that one patient achieves a partial response, and stable disease is reported in 15 patients with lasting for 6.5 months or more in 4 patients [115].

As mentioned, ubiquitination removes the process of Ub and plays an important role in genomic instability regulation and tumorigenesis processes. Thus, several DUB inhibitors have been developed and identified as potential anticancer agents [116]. G9 is described as small molecule Usp9x inhibitor suppressing Usp9x activity [117]. G9 can inhibit NRAS mutant melanoma growth by decreasing Ets-1 protein content and NRAS expression. G9 also has a synergistic effect with PD0325901, a MEK inhibitor [60]. For specific Usp9x inhibitors such as G9 targeting two other DUBs, namely Usp24 and Usp5, more drug testing is needed [117, 118]. In addition, Spautin-1 is recognized a potent USP10/13 deubiquitinating activity antagonist. A recent study revealed that Spautin-1 plays an anti-tumor role in melanoma suppression via DNA damage by increasing ROS levels and has a synergistic effect with Cisplatin [119]. Targeting USP10/13 by Spautin-1 may be a new therapeutic strategy in melanoma patient treatment.

6. Conclusions

Melanoma has a low 5-year survival rate due to being susceptible to invasion and metastasis. Recently, growing evidence identified the critical role of ubiquitination and de-ubiquitination in malignant melanoma progression, which may be the novel targets for cancer therapy. In this article, we make a brief conclusion that the misregulated expressions of the E2 ubiquitin conjugating-enzymes, E3 ubiquitin ligases, and DUBs lead to aberrant oncogenic signaling in malignant melanoma (**Figure 1**). The ubiquitination plays a vital role in melanoma not only through ubiquitination of key transcription factors or key cell signaling but also immunological microenvironment modulation. We also make a conclusion of the target UPS components, the corresponding therapeutic drugs or potential therapeutic targets, and the molecular mechanism (**Table 1**). Understanding of ubiquitination and

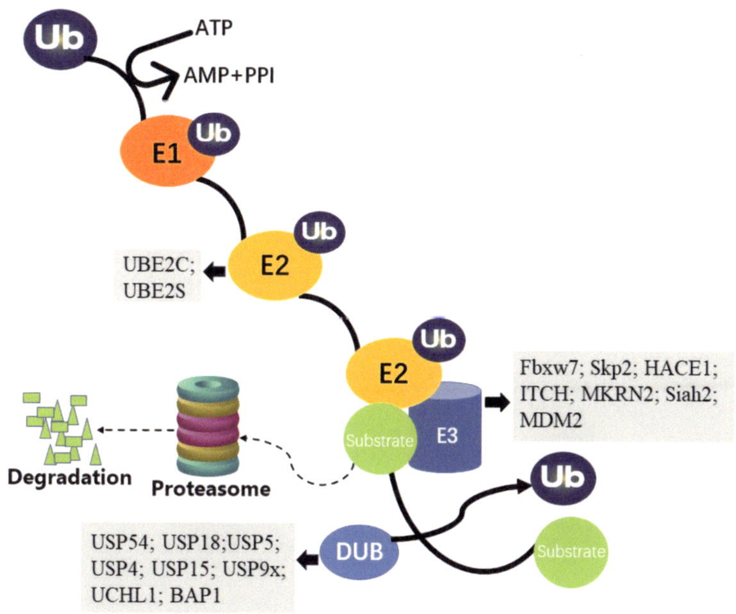

Figure 1.
Important UPS components and therapeutic targets toward melanoma pathogenesis.

Reference	UPS component	Potential therapeutic targets or drugs	Experimental model	Molecular mechanism	Clinical trial
[10]	Fbxw7	Targeting NOTCH signaling	Human and cells	Inhibiting NOTCH activation, unleashing HSF1, and activating the MAPK/ERK signaling	None
[19]	SKP2	None	Human and cells	Stabilizing the MTH1 expression via K63-linked polyubiquitination, and mediating degradation of cyclin A	None
[22, 24]	HACE1	Targeting HACE1	Human and cells	Inhibiting YAY1 signaling, and activating RAC-Family GTPases, and regulating K27 ubiquitination of FN	None
[26]	ITCH	None	Cells	Mediating BRAF polyubiquitination	None
[30, 31]	UBE2C	None	Cells	Suppressing both the level and the activity of MPF	None

Reference	UPS component	Potential therapeutic targets or drugs	Experimental model	Molecular mechanism	Clinical trial
[32]	UBE2S	Targeting UBE2S	Human, animal and cells	Catalyzing Lys11-linkage ubiquitination, and inhibiting EMT	None
[35]	MKRN2	Targeting MKRN2	Cells	Interacting with ubiquitylated P53	None
[50]	USP54	Targeting USP54	Animal and cells	Unknown	None
[52]	USP18	Targeting USP18	Animal and cells	Bing stimulated by IFN-γ, and regulating CTL CD8+ immune-cell function	None
[53]	USP5	Targeting USP5	Cells	Blocking p53 and FAS induction, and then suppressing cell cycle checkpoint and apoptosis	None
[54]	USP4	Targeting USP4	Cells	Promoting EMT	None
[55]	USP15	Targeting USP15	Animal and cells	Downregulating MDM2 expression, and increasing frequency of CD8+ effector T cell tumor-infiltrating	None
[62]	UCHL1	Targeting UCHL1	Cells	Regulating stability of MITF in human melanocytes	None
[67, 68]	BAP1	Targeting BAP1	Human, animal and cells	Unknown	None
[105, 106]	MDM2	Nutlin-3a	Human, animal and cells	Inhibiting MDM2 and cyclin B1/CDK1-phosphorylated nuclear iASPP	None
[107, 108]	Siah2	Menadione	Cells	Attenuating hypoxia and MAPK signaling	None
[59, 116]	Usp9x	G9	Cells	Decreasing Ets-1 protein content and NRAS expression, and having a synergistic effect with PD0325901	None
[111–113]	NEDD8	Pevonedistat	Human, animal and cells	Inhibiting the activity of cullin E3 ligases and then stabilizing cullin substrates	NCT01011530
[118]	USP10/13	Spautin-1	Animal and cells	Inducing DNA damage by increasing ROS levels, and having synergistic effect with Cisplatin	None

Table 1.
Summarization of the target UPS components.

de-ubiquitination mechanisms and their regulation in melanoma will help us to better understand the pathogenesis of this cancer, and develop effective therapeutic approaches, which lets us see a promising future for the application of these advancements owing to the prosperity and success of drugs targeting ubiquitination and de-ubiquitination in melanoma.

Acknowledgements

This work was supported by the National Natural Science Foundation of China (Grants No. 81772917).

Conflict of interest

The authors declare no conflict of interest.

Authors' contributions

JLZ developed the ideas and revised the manuscript. JG wrote the main manuscript.

Acronyms and abbreviations

ACT	Adoptive cell therapy
BAP1	BRCA1-associated protein-1
CRC	Colorectal cancer
DUB	Deubiquitinating enzymes
E1	The ubiquitin-activating enzyme
E2	The ubiquitin-conjugating enzyme
E3	The ubiquitin ligase
EMT	Epithelial-mesenchymal transition
Fbxw7	F-box/WD repeat-containing protein 7
FN	Fibronectin
HACE1	HECT domain and ankyrin repeat-containing E3 ubiquitin protein ligase 1
HSF1	Heat-shock factor 1
ISG15	Interferon-stimulated gene 15
MAGE-A11	Melanoma antigen-A11
MEN	Menadione
miRNAs	MicroRNAs
MITF	Microphthalmia-associated transcription factor
MKRN2	Makorin ring finger protein 2
MPF	M-phase-promoting factor
NEDD8	Neural precursor cell expressed, developmentally down-regulated8
NRF1	Nuclear Respiratory Factor 1
Skp2	S-phase kinase-associated protein 2
SUMO	Small ubiquitin-like modifiers
TLRs	Toll-like receptors
TME	Tumor microenvironment
UBE2S	Ubiquitin-conjugating enzyme E2S
UBLs	Ubiquitin-like proteins

UCHL1	Ubiquitin C-terminal hydrolase 1
UPP	Ubiquitin proteasomal pathway
UPS	Ubuiquitin-proteasome system
Usp9x	Ubiquitin specific peptidase 9, X-linked
WIP1i	WIP1 inhibitor

Author details

Jia Guo and Jianglin Guo*
Department of Dermatology, Xiangya Hospital, Central South University, Changsha, Hunan, China

*Address all correspondence to: zhangjianglin123@hotmail.com

References

[1] Garbe C, Eigentler TK, Keilholz U, Hauschild A, Kirkwood JM. Systematic review of medical treatment in melanoma: current status and future prospects. The Oncologist. 2011;**16**:5-24. DOI: 10.1634/theoncologist.2010-0190

[2] Rossi CR, Mozzillo N, Maurichi A, Pasquali S, Quaglino P, Borgognoni L, et al. The number of excised lymph nodes is associated with survival of melanoma patients with lymph node metastasis. Annals of Oncology. 2014;**25**:240-246. DOI: 10.1093/annonc/mdt510

[3] Rastrelli M, Tropea S, Rossi CR, Alaibac M. Melanoma: epidemiology, risk factors, pathogenesis, diagnosis and classification. In Vivo. 2014;**28**: 1005-1011

[4] Shmueli A, Oren M. Life, death, and ubiquitin: taming the mule. Cell. 2005;**121**:963-965. DOI: 10.1016/j.cell.2005.06.018

[5] Komander D, Rape M. The ubiquitin code. Annual Review of Biochemistry. 2012;**81**:203-229. DOI: 10.1146/annurev-biochem-060310-170328

[6] Reyes-Turcu FE, Ventii KH, Wilkinson KD. Regulation and cellular roles of ubiquitin-specific deubiquitinating enzymes. Annual Review of Biochemistry. 2009;**78**:363-397. DOI: 10.1146/annurev.biochem.78.082307.091526

[7] Grabbe C, Husnjak K, Dikic I. The spatial and temporal organization of ubiquitin networks. Nature Reviews. Molecular Cell Biology. 2011;**12**:295-307. DOI: 10.1038/nrm3099

[8] Cao J, Ge MH, Ling ZQ. Fbxw7 Tumor Suppressor: A Vital Regulator Contributes to Human Tumorigenesis. Medicine (Baltimore). 2016;**95**:e2496. DOI: 10.1097/MD.0000000000002496

[9] Yeh CH, Bellon M, Nicot C. FBXW7: a critical tumor suppressor of human cancers. Molecular Cancer. 2018;**17**:115. DOI: 10.1186/s12943-018-0857-2

[10] Aydin IT, Melamed RD, Adams SJ, Castillo-Martin M, Demir A, Bryk D, et al. FBXW7 mutations in melanoma and a new therapeutic paradigm. J Natl Cancer Inst. 2014;**106**:u107. DOI: 10.1093/jncimonographs/lgu003

[11] Kourtis N, Moubarak RS, Aranda-Orgilles B, Lui K, Aydin IT, Trimarchi T, et al. FBXW7 modulates cellular stress response and metastatic potential through HSF1 post-translational modification. Nature Cell Biology. 2015;**17**:322-332. DOI: 10.1038/ncb3121

[12] Cheng Y, Chen G, Martinka M, Ho V, Li G. Prognostic significance of Fbw7 in human melanoma and its role in cell migration. The Journal of Investigative Dermatology. 2013;**133**:1794-1802. DOI: 10.1038/jid.2013.58

[13] King R, Weilbaecher KN, McGill G, Cooley E, Mihm M, Fisher DE. Microphthalmia transcription factor. A sensitive and specific melanocyte marker for MelanomaDiagnosis. AM J PATHOL. 1999;**155**:731-738. DOI: 10.1016/S0002-9440(10)65172-3

[14] Goding CR, Arnheiter H. MITF-the first 25 years. Genes & Development. 2019;**33**:983-1007. DOI: 10.1101/gad.324657.119

[15] Abbate F, Badal B, Mendelson K, Aydin IT, Serasinghe MN, Iqbal R, et al. FBXW7 regulates a mitochondrial transcription program by modulating MITF. Pigment Cell & Melanoma Research. 2018;**31**:636-640. DOI: 10.1111/pcmr.12704

[16] Asmamaw MD, Liu Y, Zheng YC, Shi XJ, Liu HM. Skp2 in the ubiquitin-proteasome system: A comprehensive

review. Medicinal Research Reviews. 2020;**40**:1920-1949. DOI: 10.1002/med.21675

[17] Woenckhaus C, Maile S, Uffmann S, Bansemir M, Dittberner T, Poetsch M, et al. Expression of Skp2 and p27KIP1 in naevi and malignant melanoma of the skin and its relation to clinical outcome. Histology and Histopathology. 2005;**20**:501-508. DOI: 10.14670/HH-20.501.

[18] Chen G, Cheng Y, Zhang Z, Martinka M, Li G. Cytoplasmic Skp2 expression is increased in human melanoma and correlated with patient survival. PLoS One. 2011;**6**:e17578. DOI: 10.1371/journal.pone.0017578

[19] Wang JY, Liu GZ, Wilmott JS, La T, Feng YC, Yari H, et al. Skp2-Mediated Stabilization of MTH1 Promotes Survival of Melanoma Cells upon Oxidative Stress. Cancer Research. 2017;**77**:6226-6239. DOI: 10.1158/0008-5472.CAN-17-1965

[20] Su S, Chen X, Geng J, Minges JT, Grossman G, Wilson EM. Melanoma antigen-A11 regulates substrate-specificity of Skp2-mediated protein degradation. Molecular and Cellular Endocrinology. 2017;**439**:1-9. DOI: 10.1016/j.mce.2016.10.006.

[21] Anglesio MS, Evdokimova V, Melnyk N, Zhang L, Fernandez CV, Grundy PE, et al. Differential expression of a novel ankyrin containing E3 ubiquitin-protein ligase, Hace1, in sporadic Wilms' tumor versus normal kidney. Human Molecular Genetics. 2004;**13**:2061-2074. DOI: 10.1016/j.mce.2016.10.006.

[22] Zhou Z, Zhang HS, Zhang ZG, Sun HL, Liu HY, Gou XM, et al. Loss of HACE1 promotes colorectal cancer cell migration via upregulation of YAP1. Journal of Cellular Physiology. 2019;**234**:9663-9672. DOI: 10.1002/jcp.27653.

[23] Kogler M, Tortola L, Negri GL, Leopoldi A, El-Naggar AM, Mereiter S, et al. HACE1 Prevents Lung Carcinogenesis via Inhibition of RAC-Family GTPases. Cancer Research. 2020;**80**:3009-3022. DOI: 10.1158/0008-5472.CAN-19-2270

[24] El-Hachem N, Habel N, Naiken T, Bzioueche H, Cheli Y, Beranger GE, et al. Uncovering and deciphering the pro-invasive role of HACE1 in melanoma cells. Cell Death and Differentiation. 2018;**25**:2010-2022. DOI: 10.1038/s41418-018-0090-y

[25] Salah Z, Melino G, Aqeilan RI. Negative regulation of the Hippo pathway by E3 ubiquitin ligase ITCH is sufficient to promote tumorigenicity. Cancer Research. 2011;**71**:2010-2020. DOI: 10.1158/0008-5472.CAN-10-3516

[26] Yin Q, Han T, Fang B, Zhang G, Zhang C, Roberts ER, et al. K27-linked ubiquitination of BRAF by ITCH engages cytokine response to maintain MEK-ERK signaling. Nature Communications. 2019;**10**:1870. DOI: 10.1038/s41467-019-09844-0

[27] Wang S, Wu Y, Xu Y, Tang X. miR-10b promoted melanoma progression through Wnt/beta-catenin pathway by repressing ITCH expression. Gene. 2019;**710**:39-47. DOI: 10.1016/j.gene.2019.05.043

[28] Sun MX, An Q, Chen LM, Guo L. MIR-520f Regulated Itch Expression and Promoted Cell Proliferation in Human Melanoma Cells. Dose Response. 2020;**18**:710600014. DOI: 10.1177/1559325820918450

[29] Yamanaka A, Hatakeyama S, Kominami K, Kitagawa M, Matsumoto M, Nakayama K. Cell cycle-dependent expression of mammalian E2-C regulated by the anaphase-promoting complex/cyclosome. Molecular Biology of the

Cell. 2000;**11**:2821-2831. DOI: 10.1091/mbc.11.8.2821

[30] Liu G, Zhao J, Pan B, Ma G, Liu L. UBE2C overexpression in melanoma and its essential role in G2/M transition. Journal of Cancer. 2019;**10**:2176-2184. DOI: 10.7150/jca.32731

[31] Liu D, Liao C, Wolgemuth DJ. A role for cyclin A1 in the activation of MPF and G2-M transition during meiosis of male germ cells in mice. Developmental Biology. 2000;**224**:388-400. DOI: 10.1006/dbio.2000.9776

[32] Wu T, Merbl Y, Huo Y, Gallop JL, Tzur A, Kirschner MW. UBE2S drives elongation of K11-linked ubiquitin chains by the anaphase-promoting complex. Proceedings of the National Academy of Sciences of the United States of America. 2010;**107**:1355-1360. DOI: 10.1073/pnas.0912802107

[33] Wang P, Li Y, Ma Y, Zhang X, Li Z, Yu W, et al. Comprehensive investigation into the role of ubiquitin conjugating enzyme E2S (UBE2S) in melanoma development. The Journal of Investigative Dermatology. 2020. DOI: 10.1016/j.jid.2020.05.113

[34] Shin C, Ito Y, Ichikawa S, Tokunaga M, Sakata-Sogawa K, Tanaka T. MKRN2 is a novel ubiquitin E3 ligase for the p65 subunit of NF-kappaB and negatively regulates inflammatory responses. Scientific Reports. 2017;**7**: 46097. DOI: 10.1038/srep46097

[35] Zhang Y, Cui N, Zheng G. Ubiquitination of P53 by E3 ligase MKRN2 promotes melanoma cell proliferation. Oncology Letters. 2020;**19**:1975-1984. DOI: 10.3892/ol.2020.11261

[36] Enchev RI, Schulman BA, Peter M. Protein neddylation: beyond cullin-RING ligases. Nature Reviews. Molecular Cell Biology. 2015;**16**:30-44. DOI: 10.1038/nrm3919

[37] Watson IR, Irwin MS, Ohh M. NEDD8 pathways in cancer, Sine Quibus Non. Cancer Cell. 2011;**19**:168-176. DOI: 10.1016/j.ccr.2011.01.002

[38] Schwechheimer C. NEDD8-its role in the regulation of Cullin-RING ligases. Current Opinion in Plant Biology. 2018;**45**:112-119. DOI: 10.1016/j.pbi.2018.05.017

[39] Watson IR, Irwin MS. Ubiquitin and ubiquitin-like modifications of the p53 family. Neoplasia. 2006;**8**:655-666. DOI: 10.1593/neo.06439

[40] Watson IR, Li BK, Roche O, Blanch A, Ohh M, Irwin MS. Chemotherapy induces NEDP1-mediated destabilization of MDM2. Oncogene. 2010;**29**:297-304. DOI: 10.1038/onc.2009.314

[41] Russell RC, Ohh M. NEDD8 acts as a 'molecular switch' defining the functional selectivity of VHL. EMBO Reports. 2008;**9**:486-491. DOI: 10.1038/embor.2008.19

[42] Brownell JE, Sintchak MD, Gavin JM, Liao H, Bruzzese FJ, Bump NJ, et al. Substrate-assisted inhibition of ubiquitin-like protein-activating enzymes: the NEDD8 E1 inhibitor MLN4924 forms a NEDD8-AMP mimetic in situ. Molecular Cell. 2010;**37**:102-111. DOI: 10.1016/j.molcel.2009.12.024

[43] Cheng F, Chen H, Zhang L, Li RH, Liu Y, Sun JF. Inhibition of the NEDD8 conjugation pathway by shRNA to UBA3, the subunit of the NEDD8-activating enzyme, suppresses the growth of melanoma cells. Asian Pacific Journal of Cancer Prevention. 2012;**13**:57-62. DOI: 10.7314/apjcp.2012.13.1.057

[44] Moschos SJ, Jukic DM, Athanassiou C, Bhargava R, Dacic S, Wang X, et al. Expression analysis of Ubc9, the single small ubiquitin-like

modifier (SUMO) E2 conjugating enzyme, in normal and malignant tissues. Human Pathology. 2010;**41**:1286-1298. DOI: 10.1016/j.humpath.2010.02.007

[45] Moschos SJ, Smith AP, Mandic M, Athanassiou C, Watson-Hurst K, Jukic DM, et al. SAGE and antibody array analysis of melanoma-infiltrated lymph nodes: identification of Ubc9 as an important molecule in advanced-stage melanomas. Oncogene. 2007;**26**:4216-4225. DOI: 10.1038/sj.onc.1210216

[46] Bertolotto C, Lesueur F, Giuliano S, Strub T, de Lichy M, Bille K, et al. A SUMOylation-defective MITF germline mutation predisposes to melanoma and renal carcinoma. Nature. 2011;**480**:94-98. DOI: 10.1038/nature10539

[47] Mustachio LM, Kawakami M, Lu Y, Rodriguez-Canales J, Mino B, Behrens C, et al. The ISG15-specific protease USP18 regulates stability of PTEN. Oncotarget. 2017;**8**:3-14. DOI: 10.18632/oncotarget.13914

[48] Zhang D, Zhang DE. Interferon-stimulated gene 15 and the protein ISGylation system. Journal of Interferon & Cytokine Research. 2011;**31**:119-130. DOI: 10.1089/jir.2010.0110

[49] Malakhov MP, Malakhova OA, Kim KI, Ritchie KJ, Zhang DE. UBP43 (USP18) specifically removes ISG15 from conjugated proteins. The Journal of Biological Chemistry. 2002;**277**:9976-9981. DOI: 10.1074/jbc.M109078200

[50] Mustachio LM, Lu Y, Kawakami M, Roszik J, Freemantle SJ, Liu X, et al. Evidence for the ISG15-Specific Deubiquitinase USP18 as an Antineoplastic Target. Cancer Research. 2018;**78**:587-592. DOI: 10.1158/0008-5472.

[51] Fraile JM, Campos-Iglesias D, Rodriguez F, Espanol Y, Freije JM. The deubiquitinase USP54 is overexpressed in colorectal cancer stem cells and promotes intestinal tumorigenesis. Oncotarget. 2016;**7**:74427-74434. DOI: 10.18632/oncotarget.12769

[52] Ritchie KJ, Hahn CS, Kim KI, Yan M, Rosario D, Li L, et al. Role of ISG15 protease UBP43 (USP18) in innate immunity to viral infection. Nature Medicine. 2004;**10**:1374-1378. DOI: 10.1038/nm1133.

[53] Hong B, Li H, Lu Y, Zhang M, Zheng Y, Qian J, et al. USP18 is crucial for IFN-gamma-mediated inhibition of B16 melanoma tumorigenesis and antitumor immunity. Molecular Cancer. 2014;**13**:132. DOI: 10.1186/1476-4598-13-132

[54] Potu H, Peterson LF, Pal A, Verhaegen M, Cao J, Talpaz M, et al. Usp5 links suppression of p53 and FAS levels in melanoma to the BRAF pathway. Oncotarget. 2014;**5**:5559-5569. DOI: 10.18632/oncotarget.2140

[55] Guo W, Ma J, Pei T, Zhao T, Guo S, Yi X, et al. Up-regulated deubiquitinase USP4 plays an oncogenic role in melanoma. Journal of Cellular and Molecular Medicine. 2018;**22**:2944-2954. DOI: 10.1111/jcmm.13603

[56] Zou Q, Jin J, Hu H, Li HS, Romano S, Xiao Y, et al. USP15 stabilizes MDM2 to mediate cancer-cell survival and inhibit antitumor T cell responses. NAT IMMUNOL. 2014;**15**:562-570. DOI: 10.1038/ni.2885

[57] Cox JL, Wilder PJ, Wuebben EL, Ouellette MM, Hollingsworth MA, Rizzino A. Context-dependent function of the deubiquitinating enzyme USP9X in pancreatic ductal adenocarcinoma. Cancer Biology & Therapy. 2014;**15**:1042-1052. DOI: 10.4161/cbt.29182

[58] Peng J, Hu Q, Liu W, He X, Cui L, Chen X, et al. USP9X expression correlates with tumor progression

and poor prognosis in esophageal squamous cell carcinoma. Diagnostic Pathology. 2013;**8**:177. DOI: 10.1186/1746-1596-8-177

[59] Perez-Mancera PA, Rust AG, van der Weyden L, Kristiansen G, Li A, Sarver AL, et al. The deubiquitinase USP9X suppresses pancreatic ductal adenocarcinoma. Nature. 2012;**486**:266-270. DOI: 10.1186/1746-1596-8-177.

[60] Potu H, Peterson LF, Kandarpa M, Pal A, Sun H, Durham A, et al. Usp9x regulates Ets-1 ubiquitination and stability to control NRAS expression and tumorigenicity in melanoma. Nature Communications. 2017;**8**:14449. DOI: 10.1038/ncomms14449

[61] Fedorenko IV, Gibney GT, Smalley KS. NRAS mutant melanoma: biological behavior and future strategies for therapeutic management. Oncogene. 2013;**32**:3009-3018. DOI: 10.1038/onc.2012.453

[62] Larsen CN, Price JS, Wilkinson KD. Substrate binding and catalysis by ubiquitin C-terminal hydrolases: identification of two active site residues. BIOCHEMISTRY-US. 1996;**35**:6735-6744. DOI: 10.1021/bi960099f

[63] Seo EY, Jin SP, Sohn KC, Park CH, Lee DH, Chung JH. UCHL1 Regulates Melanogenesis through Controlling MITF Stability in Human Melanocytes. The Journal of Investigative Dermatology. 2017;**137**:1757-1765. DOI: 10.1016/j.jid.2017.03.024

[64] Carbone M, Yang H, Pass HI, Krausz T, Testa JR, Gaudino G. BAP1 and cancer. Nature Reviews. Cancer. 2013;**13**:153-159. DOI: 10.1038/nrc3459

[65] Jensen DE, Proctor M, Marquis ST, Gardner HP, Ha SI, Chodosh LA, et al. BAP1: a novel ubiquitin hydrolase which binds to the BRCA1 RING finger and enhances BRCA1-mediated cell growth suppression. Oncogene. 1998;**16**:1097-1112. DOI: 10.1038/sj.onc.1201861

[66] Wood LD, Parsons DW, Jones S, Lin J, Sjoblom T, Leary RJ, et al. The genomic landscapes of human breast and colorectal cancers. Science. 2007;**318**:1108-1113. DOI: 10.1126/science.1145720

[67] Wiesner T, Obenauf AC, Murali R, Fried I, Griewank KG, Ulz P, et al. Germline mutations in BAP1 predispose to melanocytic tumors. Nature Genetics. 2011;**43**:1018-1021. DOI: 10.1038/ng.910

[68] Wiesner T, Obenauf AC, Murali R, Fried I, Griewank KG, Ulz P, et al. Germline mutations in BAP1 predispose to melanocytic tumors. Nature Genetics. 2011;**43**:1018-1021. DOI: 10.1038/ng.910

[69] Harbour JW, Onken MD, Roberson ED, Duan S, Cao L, Worley LA, et al. Frequent mutation of BAP1 in metastasizing uveal melanomas. Science. 2010;**330**:1410-1413. DOI: 10.1126/science.1194472

[70] O'Shea SJ, Robles-Espinoza CD, McLellan L, Harrigan J, Jacq X, Hewinson J, et al. A population-based analysis of germline BAP1 mutations in melanoma. Human Molecular Genetics. 2017;**26**:717-728. DOI: 10.1093/hmg/ddw403

[71] Liu-Smith F, Lu Y. Opposite Roles of BAP1 in Overall Survival of Uveal Melanoma and Cutaneous Melanoma. Journal of Clinical Medicine. 2020;**9**:411. DOI: 10.3390/jcm9020411

[72] Quail DF, Joyce JA. Microenvironmental regulation of tumor progression and metastasis. Nature Medicine. 2013;**19**:1423-1437. DOI: 10.1038/nm.3394

[73] Bhoj VG, Chen ZJ. Ubiquitylation in innate and adaptive immunity. Nature. 2009;**458**:430-437

[74] Liyasova MS, Ma K, Lipkowitz S. Molecular pathways: cbl proteins in tumorigenesis and antitumor immunity-opportunities for cancer treatment. Clinical Cancer Research. 2015;**21**:1789-1794. DOI: 10.1158/1078-0432.CCR-13-2490

[75] Liu Q, Zhou H, Langdon WY, Zhang J. E3 ubiquitin ligase Cbl-b in innate and adaptive immunity. Cell Cycle. 2014;**13**:1875-1884. DOI: 10.4161/cc.29213

[76] Paolino M, Choidas A, Wallner S, Pranjic B, Uribesalgo I, Loeser S, et al. The E3 ligase Cbl-b and TAM receptors regulate cancer metastasis via natural killer cells. Nature. 2014;**507**:508-512. DOI: 10.1038/nature12998

[77] Powell DJ, Dudley ME, Hogan KA, Wunderlich JR, Rosenberg SA. Adoptive transfer of vaccine-induced peripheral blood mononuclear cells to patients with metastatic melanoma following lymphodepletion. Journal of Immunology. 2006;**177**:6527-6539. DOI: 10.4049/jimmunol.177.9.6527

[78] Rosenberg SA, Restifo NP, Yang JC, Morgan RA, Dudley ME. Adoptive cell transfer: a clinical path to effective cancer immunotherapy. Nature Reviews. Cancer. 2008;**8**:299-308. DOI: 10.1038/nrc2355

[79] Jiang X, Xu J, Liu M, Xing H, Wang Z, Huang L, et al. Adoptive CD8(+) T cell therapy against cancer:Challenges and opportunities. Cancer Letters. 2019;**462**:23-32. DOI: 10.1016/j.canlet.2019.07.017

[80] Hinterleitner R, Gruber T, Pfeifhofer-Obermair C, Lutz-Nicoladoni C, Tzankov A, Schuster M, et al. Adoptive transfer of siRNA Cblb-silenced CD8+ T lymphocytes augments tumor vaccine efficacy in a B16 melanoma model. PLoS One. 2012;**7**:e44295. DOI: 10.1371/journal.pone.0044295

[81] Li CW, Lim SO, Xia W, Lee HH, Chan LC, Kuo CW, et al. Glycosylation and stabilization of programmed death ligand-1 suppresses T-cell activity. Nature Communications. 2016;**7**:12632. DOI: 10.1038/ncomms12632

[82] Meng X, Liu X, Guo X, Jiang S, Chen T, Hu Z, et al. FBXO38 mediates PD-1 ubiquitination and regulates anti-tumour immunity of T cells. Nature. 2018;**564**:130-135. DOI: 10.1038/s41586-018-0756-0

[83] Kramer OH, Stauber RH, Bug G, Hartkamp J, Knauer SK. SIAH proteins: critical roles in leukemogenesis. Leukemia. 2013;**27**:792-802. DOI: 10.1038/leu.2012.284

[84] Ma B, Cheng H, Mu C, Geng G, Zhao T, Luo Q, et al. The SIAH2-NRF1 axis spatially regulates tumor microenvironment remodeling for tumor progression. Nature Communications. 2019;**10**:1034. DOI: 10.1038/s41467-019-08618-y

[85] Kawai T, Akira S. TLR signaling. Seminars in Immunology. 2007;**19**:24-32. DOI: 10.1016/j.smim.2006.12.004

[86] Kawai T, Akira S. Toll-like receptors and their crosstalk with other innate receptors in infection and immunity. Immunity. 2011;**34**:637-650. DOI: 10.1016/j.immuni.2011.05.006

[87] Malynn BA, Ma A. A20 takes on tumors: tumor suppression by an ubiquitin-editing enzyme. The Journal of Experimental Medicine. 2009;**206**:977-980. DOI: 10.1084/jem.20090765

[88] Guo Y, Yang L, Lei S, Tan W, Long J. NEDD4 Negatively Regulates GITR via Ubiquitination in Immune Microenvironment of Melanoma. Oncotargets and Therapy. 2019;**12**:10629-10637. DOI: 10.2147/OTT.S255478

[89] Zou Q, Jin J, Hu H, Li HS, Romano S, Xiao Y, et al. USP15 stabilizes MDM2 to mediate cancer-cell survival and inhibit antitumor T cell responses. Nature Immunology. 2014;**15**:562-570. DOI: 10.1038/ni.2885

[90] Hong B, van den Heuvel AP, Prabhu VV, Zhang S, El-Deiry WS. Targeting tumor suppressor p53 for cancer therapy: strategies, challenges and opportunities. Current Drug Targets. 2014;**15**:80-89. DOI: 10.2174/1389450114666140106101412

[91] Yuan J, Luo K, Zhang L, Cheville JC, Lou Z. USP10 regulates p53 localization and stability by deubiquitinating p53. Cell. 2010;**140**:384-396. DOI: 10.1016/j.cell.2009.12.032

[92] Girnita L, Girnita A, Larsson O. Mdm2-dependent ubiquitination and degradation of the insulin-like growth factor 1 receptor. Proceedings of the National Academy of Sciences of the United States of America. 2003;**100**:8247-8252. DOI: 10.1073/pnas.1431613100

[93] Satyamoorthy K, Chehab NH, Waterman MJ, Lien MC, El-Deiry WS, Herlyn M, et al. Aberrant regulation and function of wild-type p53 in radioresistant melanoma cells. Cell Growth & Differentiation. 2000;**11**:467-474

[94] Weiss J, Schwechheimer K, Cavenee WK, Herlyn M, Arden KC. Mutation and expression of the p53 gene in malignant melanoma cell lines. International Journal of Cancer. 1993;**54**:693-699. DOI: 10.1002/ijc.2910540427

[95] Anwar A, Norris DA, Fujita M. Ubiquitin proteasomal pathway mediated degradation of p53 in melanoma. Archives of Biochemistry and Biophysics. 2011;**508**:198-203. DOI: 10.1016/j.abb.2010.12.012

[96] Shin C, Ito Y, Ichikawa S, Tokunaga M, Sakata-Sogawa K, Tanaka T. MKRN2 is a novel ubiquitin E3 ligase for the p65 subunit of NF-kappaB and negatively regulates inflammatory responses. Scientific Reports. 2017;7:46097. DOI: 10.1038/srep46097

[97] Zhang Y, Cui N, Zheng G. Ubiquitination of P53 by E3 ligase MKRN2 promotes melanoma cell proliferation. Oncology Letters. 2020;**19**:1975-1984

[98] Hsieh AL, Walton ZE, Altman BJ, Stine ZE, Dang CV. MYC and metabolism on the path to cancer. Seminars in Cell & Developmental Biology. 2015;**43**:11-21. DOI: 10.3892/ol.2020.11261

[99] Qu X, Shen L, Zheng Y, Cui Y, Feng Z, Liu F, et al. A signal transduction pathway from TGF-beta1 to SKP2 via Akt1 and c-Myc and its correlation with progression in human melanoma. The Journal of Investigative Dermatology. 2014;**134**:159-167. DOI: 10.1016/j.semcdb.2015.08.003

[100] Hoeller D, Hecker CM, Dikic I. Ubiquitin and ubiquitin-like proteins in cancer pathogenesis. Nature Reviews. Cancer. 2006;**6**:776-788. DOI: 10.1038/jid.2013.281

[101] Appel A. Drugs: More shots on target. Nature. 2011;**480**:S40-S42. DOI: 10.1038/480S40a

[102] Bertolotto C, Lesueur F, Giuliano S, Strub T, de Lichy M, Bille K, et al. A SUMOylation-defective MITF germline mutation predisposes to melanoma and renal carcinoma. Nature. 2011;**480**:94-98. DOI: 10.1038/nature10539

[103] Bielskiene K, Bagdoniene L, Mozuraitiene J, Kazbariene B, Janulionis E. E3 ubiquitin ligases as drug targets and prognostic biomarkers in melanoma. Medicina (Kaunas, Lithuania). 2015;**51**:1-9. DOI: 10.1016/j.medici.2015.01.007

[104] Hou H, Sun D, Zhang X. The role of MDM2 amplification and overexpression in therapeutic resistance of malignant tumors. Cancer Cell International. 2019;**19**:216. DOI: 10.1186/s12935-019-0937-4

[105] Pettersson S, Sczaniecka M, McLaren L, Russell F, Gladstone K, Hupp T, et al. Non-degradative ubiquitination of the Notch1 receptor by the E3 ligase MDM2 activates the Notch signalling pathway. The Biochemical Journal. 2013;**450**:523-536. DOI: 10.1042/BJ20121249

[106] Lu M, Breyssens H, Salter V, Zhong S, Hu Y, Baer C, et al. Restoring p53 function in human melanoma cells by inhibiting MDM2 and cyclin B1/CDK1-phosphorylated nuclear iASPP. Cancer Cell. 2013;**23**:618-633. DOI: 10.1016/j.ccr.2013.03.013

[107] Ma J, Guo W, Li C. Ubiquitination in melanoma pathogenesis and treatment. Cancer Medicine. 2017;**6**:1362-1377. DOI: 10.1016/j.ccr.2013.03.013.

[108] Qi J, Nakayama K, Gaitonde S, Goydos JS, Krajewski S, Eroshkin A, et al. The ubiquitin ligase Siah2 regulates tumorigenesis and metastasis by HIF-dependent and -independent pathways. Proceedings of the National Academy of Sciences of the United States of America. 2008;**105**:16713-16718. DOI: 10.1073/pnas.0804063105

[109] Shah M, Stebbins JL, Dewing A, Qi J, Pellecchia M, Ronai ZA. Inhibition of Siah2 ubiquitin ligase by vitamin K3 (menadione) attenuates hypoxia and MAPK signaling and blocks melanoma tumorigenesis. Pigment Cell & Melanoma Research. 2009;**22**:799-808. DOI: 10.1073/pnas.0804063105.

[110] Petroski MD, Deshaies RJ. Function and regulation of cullin-RING ubiquitin ligases. Nature Reviews. Molecular Cell Biology. 2005;**6**:9-20. DOI: 10.1038/nrm1547

[111] Soucy TA, Dick LR, Smith PG, Milhollen MA, Brownell JE. The NEDD8 Conjugation Pathway and Its Relevance in Cancer Biology and Therapy. Genes & Cancer. 2010;**1**:708-716. DOI: 10.1177/1947601910382898

[112] Soucy TA, Smith PG, Milhollen MA, Berger AJ, Gavin JM, Adhikari S, et al. An inhibitor of NEDD8-activating enzyme as a new approach to treat cancer. Nature. 2009;**458**:732-736. DOI: 10.1038/nature07884

[113] Luo Z, Yu G, Lee HW, Li L, Wang L, Yang D, et al. The Nedd8-activating enzyme inhibitor MLN4924 induces autophagy and apoptosis to suppress liver cancer cell growth. Cancer Research. 2012;**72**:3360-3371. DOI: 10.1158/0008-5472.CAN-12-0388

[114] Sarantopoulos J, Shapiro GI, Cohen RB, Clark JW, Kauh JS, Weiss GJ, et al. Phase I Study of the Investigational NEDD8-Activating Enzyme Inhibitor Pevonedistat (TAK-924/MLN4924) in Patients with Advanced Solid Tumors. Clinical Cancer Research. 2016;**22**:847-857. DOI: 10.1158/1078-0432.CCR-15-1338

[115] Bhatia S, Pavlick AC, Boasberg P, Thompson JA, Mulligan G, Pickard MD, et al. A phase I study of the investigational NEDD8-activating enzyme inhibitor pevonedistat (TAK-924/MLN4924) in patients with metastatic melanoma. Investigational New Drugs. 2016;**34**:439-449. DOI: 10.1007/s10637-016-0348-5

[116] Harrigan JA, Jacq X, Martin NM, Jackson SP. Deubiquitylating enzymes and drug discovery: emerging opportunities. Nature Reviews. Drug Discovery. 2018;**17**:57-78. DOI: 10.1038/nrd.2017.152

[117] Peterson LF, Sun H, Liu Y, Potu H, Kandarpa M, Ermann M, et al. Targeting deubiquitinase activity with

a novel small-molecule inhibitor as therapy for B-cell malignancies. Blood. 2015;**125**:3588-3597. DOI: 10.1182/blood-2014-10-605584

[118] Potu H, Peterson LF, Pal A, Verhaegen M, Cao J, Talpaz M, et al. Usp5 links suppression of p53 and FAS levels in melanoma to the BRAF pathway. Oncotarget. 2014;**5**:5559-5569. DOI: 10.18632/oncotarget.2140

[119] Guo J, Zhang J, Liang L, Liu N, Qi M, Zhao S, et al. Potent USP10/13 antagonist spautin-1 suppresses melanoma growth via ROS-mediated DNA damage and exhibits synergy with cisplatin. Journal of Cellular and Molecular Medicine. 2020;**24**:4324-4340. DOI: 10.1111/jcmm.15093

Chapter 5

New Discoveries on the Roles of "Other" HECT E3 Ubiquitin Ligases in Disease Development

Emma I. Kane and Donald E. Spratt

Abstract

HECT E3 ubiquitin ligases selectively recognize, bind, and ubiquitylate their substrate proteins to target them for 26S proteasomal degradation. There is increasing evidence that HECT E3 ubiquitin ligase dysfunction due to misfolding and/or the gene encoding the protein being mutated is responsible for the development of different diseases. Apart from the more prominent and well-characterized E6AP and members of the NEDD4 family, new studies have begun to reveal how other members of the HECT E3 ubiquitin ligase family function as well as their links to disease and developmental disorders. This chapter provides a comprehensive discussion on the more mysterious members of the HECT E3 ubiquitin ligase family and how they control intracellular processes. Specifically, AREL1, HACE1, HECTD1, HECTD4, G2E3, and TRIP12 will be examined as these enzymes have recently been identified as contributors to disease development.

Keywords: apoptosis, AREL1, cancer, HECT E3 ubiquitin ligase, G2E3, HACE1, HECT, HECTD1, HECTD4, neurodevelopment, proteasomal degradation, TRIP12, ubiquitin, ubiquitylation

1. Introduction

1.1 HECT E3 ubiquitin ligase-dependent ubiquitylation

Ubiquitylation is an essential post-translational modification that regulates numerous intracellular processes including protein localization and trafficking, DNA damage response, immune system and viral response, apoptosis and proteolysis [1, 2]. E3 ubiquitin ligases play an important role in recognizing, binding, and covalently attaching ubiquitin to their various substrates to elicit a specific cellular response [3]. The homologous to E6AP C-terminus (HECT) E3 ubiquitin ligases are a unique subfamily that use a multistep pathway to selectively target substrate proteins for ubiquitylation [4]. HECT-dependent ubiquitylation requires the recruitment of an E2 ubiquitin conjugating enzyme charged with ubiquitin to the N-terminal lobe of the HECT domain on the E3 ligase [5, 6]. The ubiquitin cargo is then transferred from the E2 enzyme to the conserved catalytic cysteine within the C-terminal lobe of the HECT domain *via* a transthiolation reaction to form a thioester bond. The HECT E3~ubiquitin complex will then bind to a substrate and covalently attach ubiquitin on to a lysine residue of the substrate protein forming a stable isopeptide bond between the C-terminus of ubiquitin and the ε-amine of the substrate lysine [3, 5, 6]. This process can be repeated numerous times to form different polyubiquitin chain linkages with the specific HECT E3 ubiquitin ligase dictating the type(s) of ubiquitin linkages that are built [2, 7].

Chain types	Linker	Proposed function
Monoubiquitylation		
Monoubiquitylation/ multi-monoubiquitylation		Endocytosis [9] DNA damage repair [10–15] Histone regulation [10–15] Mitophagy [10–15] Protein localization [10–15] Protein interactions [10–15] Protein transportation [10–15] Transcription activation [10–15]
Polyubiquitylation		
Chain (homotypic)	M1	Innate immunity [2, 9, 16] Linear chain formation [9] NF-κB activation [9, 16] Signaling cascades [9, 16]
	K6	DNA damage response [14] NF-κB regulation [14] Mitophagy [14]
	K11	Cell cycle regulation [17] DNA damage response [18] Mitophagy [17] NF-κB activation [16] Protein degradation [17]
	K27	DNA damage response [18] Kinase activation [19] Protein degradation [20] Protein scaffolding [21] Protein trafficking [22]
	K29	DNA damage response [18] Kinase activation [19] Protein degradation [9]
	K33	DNA damage response [10–15, 18] Kinase activation [23] Post-golgi trafficking [24] T-cell signaling [23]
	K48	Protein degradation [1, 2, 25]
	K63	DNA damage response [9, 18] NF-κB activation [9, 16] Protein trafficking [9]
Chain (heterotypic; branched)	M1/K63	NF-κB activation [16]
	K11/K48	Protein degradation [26, 27]
	K29/K48	Protein degradation [26]
	K48/K63	Protein degradation [26]
	K11/K63	Endocytosis [28]

Table 1.
Ubiquitin conjugation determines the intracellular fate of a substrate protein.

1.2 Ubiquitin attachment site(s) and chain type linkages determine the fate of a substrate protein

The destiny of a ubiquitin-tagged protein is dependent on (i) the site(s) of ubiquitin attachment on the substrate, (ii) the number of ubiquitin moieties attached to the substrate (i.e. mono-, multi-mono-, or polyubiquitin), and (iii) the specific type(s) of linkages between the different ubiquitin molecules in a polyubiquitin chain (i.e. K48, K63, branched, etc.) [1, 2, 7]. Potential fates of a ubiquitin-tagged substrate include changes in cellular localization/trafficking, enhanced/inhibited protein activity, changes in protein–protein affinity/interactions, and proteolysis [1, 2, 6–8].

Differences in ubiquitin lysine linkage specificity determine the destination and/or fate of the targeted protein in the cell (**Table 1**). For example, the well-established K48-polyubiquitylation chain, heterotypic K11/K48-polyubiquitin, K29/K48-polyubiquitin monoubiquitin tagged peptides and multiple monoubiquitin tagged proteins have also been found to signal for 26S proteasomal degradation [7, 8]. K63-polyubiquitin chains signal for protein degradation through the initiation of K48/K63 polyubiquitin branch formation [5] but cannot be recognized by the 26S proteosome [26]. To date, many different varieties of ubiquitin chain types have been identified, but their distinct biological functions remain unclear.

Monoubiquitylation can occur at one site or at multiple sites (multi-monoubiquitylation) on a substrate. Polyubiquitylation can build off of a mono-ubiquitin attachment site with a specific lysine linkage (homotypic) or have multiple chains with different lysine linkages (branch) at the end of a growing ubiquitin chain (heterotypic). These modifications can also influence signaling pathways, whether it is through enhancing or inhibiting participating proteins and processes.

2. The "other" HECT E3 ubiquitin ligases: important players in disease, yet poorly understood

The HECT E3 ubiquitin ligases can be categorized into three subfamilies – NEDD4, HERC, and "other" – based on their sequence/structure similarity and domain architecture [4, 5]. Of the 28 HECT E3 ubiquitin ligases identified in humans, there are 12 "other" HECT E3 ubiquitin ligases that do not fall under the well-studied NEDD4 or HERC subfamilies. Each member of the "other" HECTs have variable N-terminal domains that are thought to be involved in protein–protein interactions and/or intracellular localization [4, 5]. Having prominent responsibilities in cellular homeostasis would leave the impression there is ample research on the HECT E3 ubiquitin ligase family as a whole, however, there remain many unanswered questions about the biological functions and mechanisms of this important E3 ligase family, particularly for members of the more mysterious "other" subfamily. With new research and discoveries becoming available, there is mounting evidence that the lesser known HECT E3 ubiquitin ligases play critical roles in regulating intracellular processes and their dysfunction have been suggested to contribute to the onset of many diseases and disorders [4, 29–32]. Here we discuss the latest discoveries on these lesser known members of the HECT E3 ubiquitin ligases and on their emerging roles in developmental and neurological abnormalities, cancers, and embryogenesis.

2.1 AREL1, a key regulator of apoptosis and potential oncogenic drug target

Apoptosis resistant E3 ubiquitin protein ligase 1(AREL1; 823 residues) is a cytosolic enzyme responsible for regulating apoptosis through the inhibition of proapoptotic proteins *via* ubiquitylation [33]. AREL1 contains two immunoglobin-like folds (IGF) near its N-terminus that potentially mediate substrate binding and recognition (**Figure 1**) [35, 36]. IGF domains assemble into a sandwich-like form consisting of antiparallel β-strands that allow for protein–protein interactions [36, 37].

Apoptosis (aka programmed cell death) is an important and highly regulated biological process that occurs during early embryonic development and the immune response [38, 39]. Once apoptosis is initiated the cell is committed to die, which is mediated by a caspase cascade [40]. Intrinsic apoptosis is turned on by the release of intermembrane mitochondrial proteins when cells are under oxidative stress [38, 39, 41]. In contrast, the extrinsic apoptotic pathway is activated by extracellular signaling at the cell membrane leading to the formation of the death-inducing signaling complex (DISC) [42–44].

AREL1 was first identified in 2013 and was immediately recognized as an oncogenic target due to its inhibitory role in apoptosis [33]. Identified substrates for AREL1 include second mitochondrial activator of caspase (SMAC), HtrA serine peptidase 2 (HtrA2) and septin 4 (ARTS), which are known antagonists of inhibitor of apoptosis proteins (IAPs) [45]. Studies have shown that AREL1 can build K48 and K63 polyubiquitin chains to target substrates for proteolysis, as well as atypical K33 polyubiquitin chains whose biological function is still being clarified [35, 46]. Various IAPs, including SMAC, HtrA1, and ARTS, are released from the mitochondrial intermembrane into the cytosol when the cell is triggered or stressed. AREL1 inhibits apoptosis by ubiquitylating these IAP antagonists with K48-linked polyubiquitin chains targeting the IAPs for proteasomal degradation [33].

The induction of apoptosis is thought to require the release of numerous IAPs in the cytosol to allow different signaling pathways to initiate apoptosis depending on the cell's specific stress. For example, the release of SMAC into the cytosol allows it to bind cellular inhibitor of apoptosis protein 1/2 (cIAP1/2), which then targets cIAP1/2 for proteasomal degradation to initiate apoptosis. However, when AREL1 is present, SMAC is ubiquitylated by AREL1 and degraded, thus blocking SMAC-cIAP1/2 complex formation enabling cell survival [33]. Many cancer therapies are interested in specifically turning on apoptosis through IAPs in cancer cells [47–49],

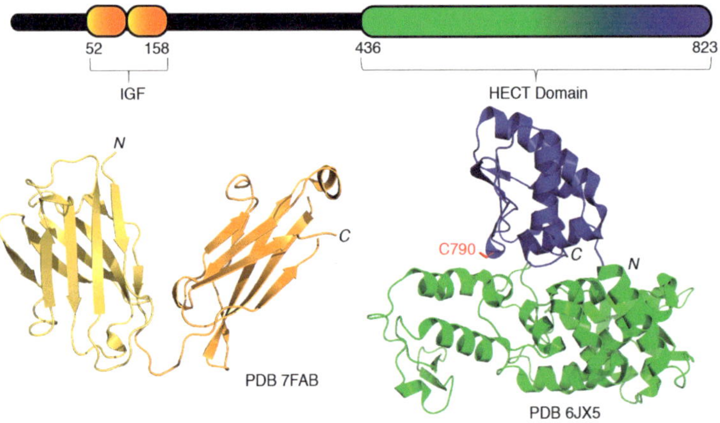

Figure 1.
AREL1 domain architecture. AREL1 contains a putative immunoglobulin fold domain (IGF, residues 52-158) and the canonical HECT domain (436-823) as annotated on UniProt and InterPro. Representative crystal structures of an IGF fold (human IGF FAB in yellow/orange; PDB 7FAB [34]), which is suggested to mediate AREL1 substrate binding and recognition, while the AREL1 HECT domain (HECT$^{N\text{-}lobe}$ in green, HECT$^{C\text{-}lobe}$ in blue, catalytic cysteine C790 in red; PDB 6JX5 [35]) is required for ubiquitylation activity. Structures were visualized using PyMol.

thus AREL1 could prove to be a novel enzyme in drug development. Continued studies on AREL1's mechanisms in controlling cell death are warranted.

2.2 HACE1, a prominent tumor suppressor with dual function

First identified in 2004, HECT domain and ankyrin repeat containing E3 ubiquitin protein ligase 1 (HACE1; 909 residues) has been shown to take part in various cellular processes. For example, HACE1 is best known as a tumor suppressor as altered HACE1 expression levels have been observed in various cancers including colorectal, breast, liver, kidney, osteosarcoma, lymphoma and gastric cancer [50–54]. HACE1 contains six ankyrin repeats near its N-terminus that likely take part in HACE1-substrate recognition and protein–protein interactions (**Figure 2**). While it is not yet fully understood how the ankyrin repeats support HACE1 function, ankyrin repeats in other proteins have been shown to instigate the development of a wide array of diseases including cancer [56]. HACE1 also supports Golgi membrane biogenesis during cell division by ubiquitylating members of the Ras-related superfamily of small G proteins [57].

Studies have shown that HACE1 expression levels are altered when comparing normal and cancerous tissue. Specifically, in the Wilms' tumor cell line, HACE1 expression was essentially nonexistent, whereas in other cancer cell lines expression levels were lower than average [50]. This study concluded that HACE1 was essential in repressing cancer development as the lowered expression levels of HACE1 were not mutation dependent. Low expression levels of HACE1 have also been observed in other cancer cell lines. For instance, it was found that the methylation of the HACE1 promoter resulted in decreased HACE1 expression in liver cancer cells, which in turn decreased HACE1's ability to ubiquitylate its identified substrates optineurin (OPTN) and microtubule-associated proteins 1A/1B light chain 3B protein [53]. Many different substrates of HACE1 have been identified to date (summarized in [4]), including β2-adrenergic receptor (ß2AR) [58], OPTN [59], retinoic acid receptor beta (RAR-β) [57], tumor necrosis factor receptor-2 (TNFR2) [60], and various Ras-related

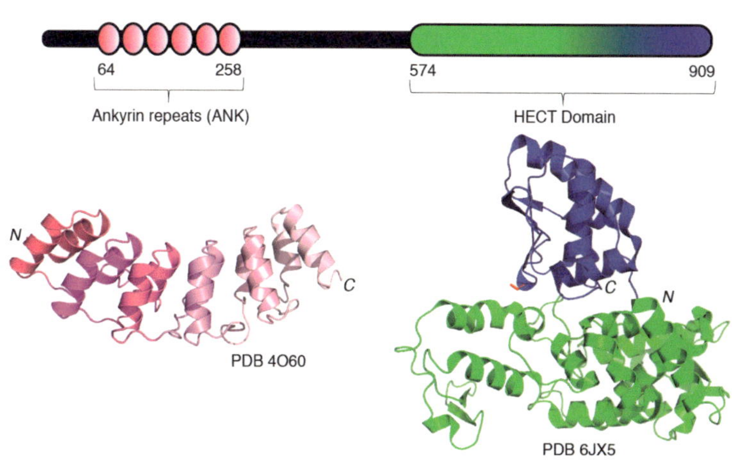

Figure 2.
Predicted domain architecture of HACE1. HACE1 contains six putative ankyrin-repeats (ANK; residues 64-258) and a HECT domain (574-909) as annotated on UniProt and InterPro. Representative crystal structures of an ANK repeat fold (in shades of pink, PDB 4O60 [55]), which is likely involved in HACE1 substrate binding and recognition, and a HECT domain (HECTN-lobe in green, HECTC-lobe in blue; PDB 6JX5 [35]) found at HACE1's C-terminus that is required of ubiquitylation activity. Structures were visualized using PyMol.

proteins [57, 58, 61–65]. Expanded studies on how HACE1 binds and recognizes its substrates are needed to further clarify the role of HACE1 in cancer development.

HACE1 also plays an essential role in neurodevelopment as it was recently shown to be involved in the spastic paraplegia and psychomotor retardation with or without seizures (SPPRS) phenotype [66]. HACE1 also has cardiac protection function during hemodynamic stress when it was shown in mice with *HACE1* deficiency, their susceptibility to accelerated heart failure greatly increased [67]. This suggests that HACE1 has a critical role in protecting the heart from various stresses, thus making it a potential cardiac drug target.

2.3 HECTD1, an important regulator in neurodevelopment

HECT domain containing E3 ubiquitin protein ligase 1 (HECTD1; 2610 residues) was discovered in 2007 as a novel and important regulator of neurodevelopment [68]. HECTD1 has similar domain architecture to HACE1 with four ankyrin repeats near its N-terminus and a C-terminal HECT domain (**Figure 3**). HECTD1 plays an important role in pulmonary fibrosis during endothelial-mesenchymal transition (EndMT) with reports of increased circular RNA HECTD1 (circHECTD1) transcription, which causes decreased HECTD1 protein expression in lung tissue [72, 73]. Elevated circHECTD1 gene expression has also been found in patients with acute ischemic stroke (AIS) [74]. HECTD1 also contains a Smad4 activation SAD1/UNC (SUN) domain and a mind bomb (MIB) domain, with each having unique roles in intracellular signaling due to Smad-DNA complex formation and cellular interactions through the Notch pathway, respectively [75, 76].

HECTD1 supports fetal growth and proper placenta development. Specifically, HECTD1 aids in the development of the labyrinthine and junctional zones of the placenta, regions where the fetus acquires nutrients and disposes of waste, as well as a bilayer between the labyrinthine and decidual cells, respectively [77, 78]. HECTD1 ensures the proper size development of the labyrinthine, yet the mechanisms to ensure this are still not fully understood. Mutations within HECTD1 lead to the onset of irregular labyrinthine development, which in turn depletes nutrients for the fetus. Fetal fatality can occur without proper maintenance of these placental regions, suggesting

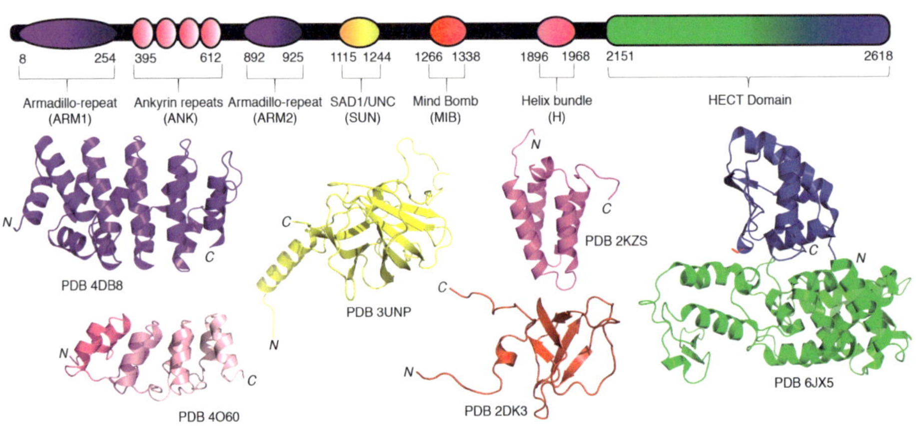

Figure 3.
HECTD1 domain architecture. HECTD1 contains putative protein-protein interaction domains including two armadillo-repeat containing domains (ARM1, residues 8-254; ARM2 residues 892-925 in purple; PDB 4DB8 [69]), four ankyrin-repeats (residues 395-612 in shades of pink; PDB 4O60 [56]), a SAD1/UNC domain (SUN, residues 1115-1244 in yellow; PDB 3UNP [70]), a mind bomb domain (MIB, residues 1266–1338 in red; PDB 2DK3), a helix-bundle domain (H, residues 1896–1968 in magenta; PDB 2KZS [71]) and a HECT domain (HECT$^{N\text{-}lobe}$ in green, HECT$^{C\text{-}lobe}$ in blue; PDB 6JX5 [35]). Domain boundaries are denoted according to UniProt and InterPro. Structures were visualized using PyMol.

that HECTD1 expression is essential for the proper development and survival of fetuses *in utero*. HECTD1 also plays a role in proper neural tube closure. Anencephaly occurs when the neural tube does not close properly, which has been linked to HECTD1 control of heat shock protein 90 (Hsp90) levels [79]. When Hsp90 secretion levels are not properly regulated by HECTD1 ubiquitylation, abnormal neural tube development can occur. The continued examination of this important enzyme will hopefully clarify the molecular basis for HECTD1's role in neurodevelopment.

2.4 HECTD4, a genetically linked precursor to cancer and cardiovascular disease

HECT domain containing E3 ubiquitin protein ligase 4 (HECTD4; 3996 residues) was recently discovered in 2014. A pleiotropic gene screen showed that there were links between metabolic syndromes and inflammation, specifically with single nucleotide polymorphisms (SNPs) in the *HECTD4* gene [80]. Since 2014, HECTD4 has also been found to be associated with diabetes, hypertension and cardiovascular disease, lung adenocarcinoma, urothelial carcinoma and ovarian endometriosis [81–84].

Being one of the larger enzymes within the "other" HECT E3 ubiquitin ligase subfamily, it is intriguing that there have been no putative domains annotated for HECTD4 except the C-terminal HECT domain (residues 3627–3996). There are likely different domains located in the N-terminal region of the HECTD4 protein that need to be identified and functionally examined.

Genetic screening has identified various mutations in *HECTD4* in cancer cells. For example, *HECTD4* was recently identified as one of nine genes that correlated with the onset of lung adenocarcinoma [83]. Tumor genetic screens have revealed in patients with urothelial carcinoma in the bladder (UCB) that mutations in *HECTD4*, Fibrillin-3 Precursor (*FBN3*) and Citron Rho-Interacting Kinase (*CIT*) were correlated to UCB disease progression [81]. *HECTD4* may also be linked to the development of ovarian endometriosis (OEM) [82]. However, future studies are still needed to verify this relationship.

Already having various genetic links to cancers and cardiovascular disease, HECTD4 deserves more attention by the research community to further clarify its biological and functional roles in the cell. Currently very little is known about HECTD4, therefore it will be imperative to first identify potential similarities in protein sequence and/or domain architecture. To better clarify HECTD4's role in ubiquitin biology, it will also be important to discover HECTD4 substrates and annotate the sites of HECTD4-dependent ubiquitylation to answer how this mysterious HECT E3 ubiquitin ligase contributes to disease development.

2.5 G2E3, a unique multifunctional HECT E3 ubiquitin ligase with RING-like features

G2/M-phase specific E3 ubiquitin protein ligase (G2E3; 706 residues) was first identified in 2006 and named for its role during the G2/M phase of cell division and for having a conserved C-terminal HECT domain [85]. Knockout studies of G2E3 in mice demonstrated that this enzyme is essential in preventing apoptotic cell death during early embryonic development [86]. Expression levels in G2E3 were also observed to increase during early embryogenesis, specifically during central nervous system development. This enzyme is also implicated in cell cycle progression and DNA damage response [85, 87, 88].

G2E3 contains three plant homeodomain (PHD)-type zinc finger repeats, a domain typically known to bind to modified histones and act as epigenetic readers [89], making it the only known HECT E3 ubiquitin ligase to possess "RING"-like

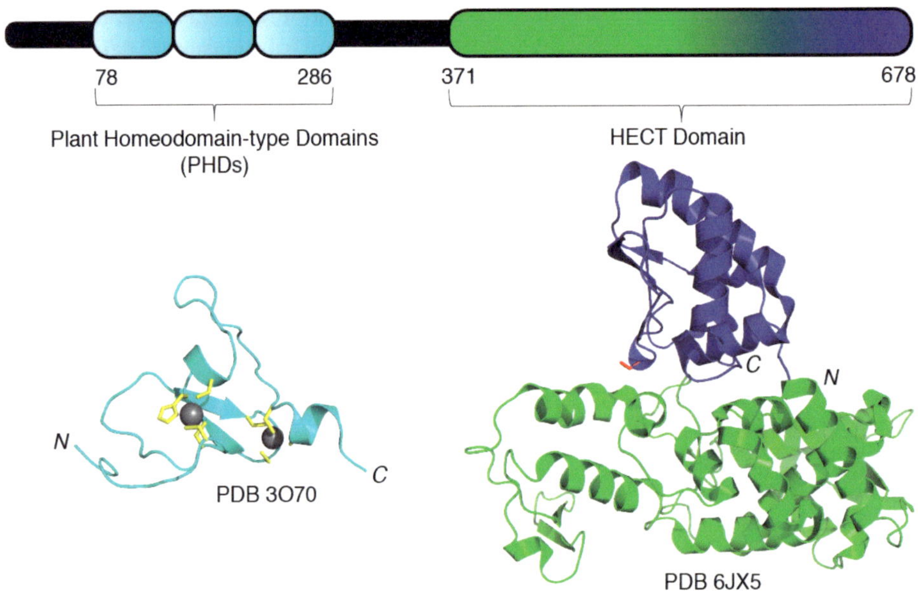

Figure 4.
Domain architecture of G2E3. G2E3 contains three putative plant homeodomain (PHD)-type domains (residues 78-128) and a unique HECT-like domain (371-678) as annotated on UniProt and InterPro. Representative crystal structures of a PHD domain (human PHD finger protein 13 in cyan with Zn-coordinating residues in yellow; PDB 3O70 [90]), which is suggested to interact with DNA and/or proteins in the nucleus, while the G2E3 HECT domain (HECT^{N-lobe} in green, HECT^{C-lobe} in blue; PDB 6JX5 [35]) may have lost its ability to catalyze the transfer of ubiquitin. Structures were visualized using PyMol.

characteristics (**Figure 4**). G2E3 is primarily found within the nucleus due to its N-terminal nucleolar localization signal sequence, while the PHD domains have been suggested to cause the translocation of G2E3 to the cytoplasm [85, 86]. Previous biochemical studies showed that the ubiquitin ligase activity of G2E3 was exclusively found in two of the putative N-terminal PHD domains while the C-terminal HECT domain of G2E3 had apparently lost its ubiquitylation activity [86]. Since the PHD domains of G2E3 appear to be capable of recruiting E2 enzymes to build K48-linked polyubiquitin chains [86], this suggests that the HECT domain of G2E3 may have become vestigial through evolutionary pressure. Intriguingly, G2E3 has aspects of the RING and HECT E3 ubiquitin ligase families, analogous to members of the RING-between-RING (aka RING-BRcat-Rcat) E3 ubiquitin ligases that includes parkin and HOIL-interacting protein (HOIP) of the LUBAC complex [91, 92].

G2E3 was recently identified as a potential drug target to increase the efficacy of chemotherapy drugs, specifically with Cisplatin [88]. Since Cisplatin is the most common chemotherapy drug, much research has been dedicated to increasing Cisplatin's ability to specifically trigger the DNA damage response in cancer cells to initiate apoptosis while limiting its exposure time and prescribed duration for patients [93]. Clearly, G2E3 is an important nuclear protein whose mechanism is currently unresolved. Further studies are needed to clarify how this divergent HECT-domain containing E3 ubiquitin ligase works in the cell.

2.6 TRIP12, the multifunctioning E3 ubiquitin ligase essential for embryogenesis and DNA damage repair

Thyroid hormone receptor interactor 12 (TRIP12; 2040 residues), was first identified in 2001 for containing a unique tryptophan-tryptophan-glutamate (WWE) domain that is predicted to be involved in ubiquitylation and ADP-ribosylation [94] (**Figure 5**). It also contains two N-terminal ARM domains, similar to HECTD1, and

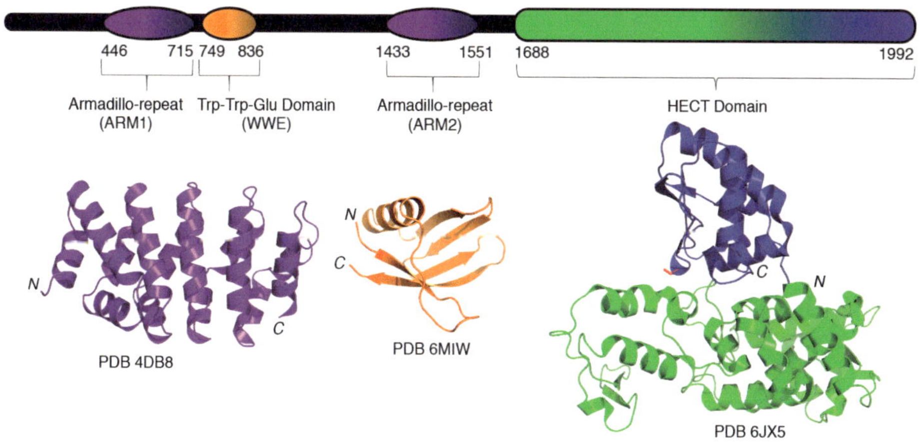

Figure 5.
TRIP12 domain architecture. TRIP12 contains two putative armadillo-repeat containing domains (ARM1, residues 437-713; ARM2, residues 826-938), a tryptophan-tryptophan-glutamate (WWE)-domain (residues 749-836) and a conserved HECT domain (1885-1992) as annotated on UniProt and InterPro. Representative crystal structures of an ARM domain (in purple; PDB 4DB8 [69]) and WWE domain (in orange; PDB 6MIW), both of which are suggested to be involved in protein-protein and substrate interactions, as well as a HECT domain (HECT$^{N\text{-}lobe}$ in green, HECT$^{C\text{-}lobe}$ in blue; PDB 6JX5 [35]) required for ubiquitylation activity. Structures were visualized using PyMol.

a conserved HECT domain at its C-terminus. TRIP12 is a novel HECT E3 ubiquitin ligase that has been shown to take part in various cellular pathways and processes including embryogenesis, DNA damage response and the neddylation pathway [95–97]. It has been reported that TRIP12 preferentially builds mono- as well as K48 and K63 polyubiquitin chains to tag its substrates for degradation and for DNA damage site recruitment, respectively [96].

TRIP12 has been shown to be directly and/or indirectly involved in cancer progression. For example, TRIP12 may serve as an oncogenic drug target for patients with acute myeloid leukemia (AML) by blocking a TRIP12 alternative splicing event, specifically excising exon3 from the mature TRIP12 mRNA [98]. TRIP12 also targets pancreas transcription factor 1a (PTF1a) for proteasomal degradation, a protein essential for pancreatic cancer development [99]. TRIP12 forms a ternary complex with deubiquitylase ubiquitin-specific protease 7 (USP7) that aids in hepatocellular carcinoma (HCC) proliferation; when USP7 expression levels are heightened, TRIP12 cannot tag ARF tumor suppressor (p14ARF) for ubiquitylation [100]. Furthermore, TRIP12 is associated with human papillomavirus (HPV)-positive head and neck squamous cell carcinoma (HNSCC) due to its mediation of p16-related radiation efficacy [101].

Members of the HECT E3 ubiquitin ligase family play important roles in neurodevelopment and their malfunction may be causative in different neurological diseases and disorders (reviewed in [4]). Recent genetic screens have been looking to identify genetic markers for autism spectrum disorder (ASD) and intellectual disability (ID). Interestingly, a *de novo* mutation in *TRIP12* was found in patients with or without ASD and displaying characteristics of ID [102]. Further studies to clarify the specific mechanism(s) for how mutations in the *TRIP12* gene contribute to ASD and ID phenotypes are needed.

3. Conclusion

Although much research has and continues to be performed for E6AP and members of the NEDD4 family, greater attention on the mysterious "other"

HECT E3 ubiquitin ligases is warranted due to their emerging involvement in various diseases and neurological disorders. Combining genetic, cell biology, biochemical, and biophysical approaches to study these unique HECT E3 ubiquitin ligases will help to decipher their specific roles and/or functions in the cell as well as potentially aid in novel therapy development to treat rare conditions caused by the dysfunction of these lesser known members of the HECT E3 ubiquitin ligase family.

Acknowledgements

This work was supported by the National Institutes of Health (R15GM126432 to D.E.S.) and start-up funds from Clark University (to D.E.S.).

Conflict of interest

The authors declare that there is no conflict of interest regarding the publication of this chapter.

Abbreviations

AIS	acute ischemic stroke
AML	acute myeloid leukemia
ANK	ankyrin repeat
AREL1	apoptosis resistant E3 ubiquitin protein ligase 1
ARM	armadillo-repeat domain
ARTS	septin 4
ASD	autism spectrum disorder
β2AR	adrenergic receptor β2AR
cIAP1/2	cellular inhibitor of apoptosis protein 1/2
circHECTD1	circular RNA HECTD1
CIT	citron rho-interacting kinase
EndMT	endothelial-mesenchymal transition
FBN3	fibrillin-3 precursor
G2E3	G2/M-phase specific E3 ubiquitin protein ligase
H	helix-bundle domain
HACE1	HECT domain and ankyrin repeat containing E3 ubiquitin protein ligase 1
HCC	hepatocellular carcinoma
HECT	hmologous to E6AP C-terminus
HECTD1	HECT domain containing E3 ubiquitin protein ligase 1
HECTD4	HECT domain containing E3 ubiquitin protein ligase 4
HERC	HECT and RLD domain-containing
HNSCC	head and neck squamous cell carcinoma
HOIP	HOIL-interacting protein
HPV	human papillomavirus
Hsp90	heat shock protein 90
HtrA2	HtrA serine peptidase 2
ID	intellectual disability
IGF	immunoglobulin-like fold
MIB	mind bomb domain

NEDD4	neuronal precursor cell-expressed developmentally down-regulated 4
SMAC	second mitochondrial activator of caspase
OEM	ovarian endometriosis
OPTN	optineurin
p14ARF	ARF tumor suppressor
PHD	plant homeodomain-type zinc finger
PTF1a	pancreas transcription factor 1a
RAR-β	retinoic acid receptor beta
RING	really interesting new gene
TNFR2	tumor necrosis factor receptor-2
SNPs	single nucleotide polymorphisms
SPPRS	spastic paraplegia and psychomotor retardation with or without seizures
SUN	Smad4 activation SAD1/UNC domain
TRIP12	thyroid hormone receptor interactor 12
UCB	urothelial carcinoma in the bladder
USP7	deubiquitylase ubiquitin-specific protease 7
WWE	tryptophan-tryptophan-glutamate domain

Author details

Emma I. Kane and Donald E. Spratt*
Clark University, Worcester, Massachusetts, United States

*Address all correspondence to: dspratt@clarku.edu

References

[1] Hershko A, Ciechanover A. The ubiquitin system. Annual Review of Biochemistry. 1998;**67**:425-479

[2] Komander D, Rape M. The ubiquitin code. Annual Review of Biochemistry. 2012;**81**:203-229

[3] Morreale FE, Walden H. Types of ubiquitin ligases. Cell. 2016;**165**(1):248-e1

[4] Wang Y, Argiles-Castillo D, Kane EI, Zhou A, Spratt DE. HECT E3 ubiquitin ligases: Emerging insights into their biological roles and disease relevance. Journal of Cell Science. 2020. in press

[5] Lorenz S. Structural mechanisms of HECT-type ubiquitin ligases. Biological Chemistry. 2018;**399**(2):127-145

[6] Metzger MB, Hristova VA, Weissman AM. HECT and RING finger families of E3 ubiquitin ligases at a glance. Journal of Cell Science. 2012;**125**(Pt 3):531-537

[7] Swatek KN, Komander D. Ubiquitin modifications. Cell Research. 2016;**26**(4):399-422

[8] Grice GL, Nathan JA. The recognition of ubiquitinated proteins by the proteasome. Cellular and Molecular Life Sciences. 2016;**73**(18):3497-3506

[9] Pickart CM, Fushman D. Polyubiquitin chains: Polymeric protein signals. Current Opinion in Chemical Biology. 2004;**8**(6):610-616

[10] Bergink S, Jentsch S. Principles of ubiquitin and SUMO modifications in DNA repair. Nature. 2009;**458**(7237): 461-467

[11] Durcan TM, Tang MY, Perusse JR, Dashti EA, Aguileta MA, McLelland GL, et al. USP8 regulates mitophagy by removing K6-linked ubiquitin conjugates from parkin. The EMBO Journal. 2014;**33**(21):2473-2491

[12] Haglund K, Di Fiore PP, Dikic I. Distinct monoubiquitin signals in receptor endocytosis. Trends in Biochemical Sciences. 2003;**28**(11): 598-603

[13] Hicke L. Protein regulation by monoubiquitin. Nature Reviews. Molecular Cell Biology. 2001;**2**(3):195-201

[14] Passmore LA, Barford D. Getting into position: The catalytic mechanisms of protein ubiquitylation. The Biochemical Journal. 2004;**379**(Pt 3):513-525

[15] Paul A, Wang B. RNF8- and Ube2S-dependent ubiquitin lysine 11-linkage modification in response to DNA damage. Molecular Cell. 2017;**66**(4):458-472.e5

[16] Iwai K. Diverse ubiquitin signaling in NF-kappaB activation. Trends in Cell Biology. 2012;**22**(7):355-364

[17] Wickliffe KE, Williamson A, Meyer HJ, Kelly A, Rape M. K11-linked ubiquitin chains as novel regulators of cell division. Trends in Cell Biology. 2011;**21**(11):656-663

[18] Liu P, Gan W, Su S, Hauenstein AV, Fu TM, Brasher B, et al. K63-linked polyubiquitin chains bind to DNA to facilitate DNA damage repair. Science Signaling. 2018;**11**(533)

[19] Singh R, Karri D, Shen H, Shao J, Dasgupta S, Huang S, et al. TRAF4-mediated ubiquitination of NGF receptor TrkA regulates prostate cancer metastasis. The Journal of Clinical Investigation. 2018;**128**(7):3129-3143

[20] Birsa N, Norkett R, Wauer T, Mevissen TE, Wu HC, Foltynie T, et al. Lysine 27 ubiquitination of

the mitochondrial transport protein Miro is dependent on serine 65 of the Parkin ubiquitin ligase. The Journal of Biological Chemistry. 2014;**289**(21):14569-14582

[21] Wang Q, Liu X, Cui Y, Tang Y, Chen W, Li S, et al. The E3 ubiquitin ligase AMFR and INSIG1 bridge the activation of TBK1 kinase by modifying the adaptor STING. Immunity. 2014;**41**(6):919-933

[22] Castaneda CA, Dixon EK, Walker O, Chaturvedi A, Nakasone MA, Curtis JE, et al. Linkage via K27 bestows ubiquitin chains with unique properties among polyubiquitins. Structure. 2016;**24**(3): 423-436

[23] Huang H, Jeon MS, Liao L, Yang C, Elly C, Yates JR 3rd, et al. K33-linked polyubiquitination of T cell receptor-zeta regulates proteolysis-independent T cell signaling. Immunity. 2010;**33**(1):60-70

[24] Yuan WC, Lee YR, Lin SY, Chang LY, Tan YP, Hung CC, et al. K33-linked polyubiquitination of coronin 7 by Cul3-KLHL20 ubiquitin E3 ligase regulates protein trafficking. Molecular Cell. 2014;**54**(4):586-600

[25] Thrower JS, Hoffman L, Rechsteiner M, Pickart CM. Recognition of the polyubiquitin proteolytic signal. The EMBO Journal. 2000;**19**(1):94-102

[26] Ohtake F, Tsuchiya H, Saeki Y, Tanaka K. K63 ubiquitylation triggers proteasomal degradation by seeding branched ubiquitin chains. Proceedings of the National Academy of Sciences of the United States of America. 2018;**115**(7):E1401-E1408

[27] Meyer HJ, Rape M. Enhanced protein degradation by branched ubiquitin chains. Cell. 2014;**157**(4):910-921

[28] Boname JM, Thomas M, Stagg HR, Xu P, Peng J, Lehner PJ. Efficient internalization of MHC I requires lysine-11 and lysine-63 mixed linkage polyubiquitin chains. Traffic. 2010;**11**(2):210-220

[29] Bernassola F, Chillemi G, Melino G. HECT-type E3 ubiquitin ligases in cancer. Trends in Biochemical Sciences. 2019;**44**(12):1057-1075

[30] Bielskiene K, Bagdoniene L, Mozuraitiene J, Kazbariene B, Janulionis E. E3 ubiquitin ligases as drug targets and prognostic biomarkers in melanoma. Medicina (Kaunas, Lithuania). 2015;**51**(1):1-9

[31] Fajner V, Maspero E, Polo S. Targeting HECT-type E3 ligases - insights from catalysis, regulation and inhibitors. FEBS Letters. 2017;**591**(17):2636-2647

[32] Weber J, Polo S, Maspero E. HECT E3 ligases: A tale with multiple facets. Frontiers in Physiology. 2019;**10**:370

[33] Kim JB, Kim SY, Kim BM, Lee H, Kim I, Yun J, et al. Identification of a novel anti-apoptotic E3 ubiquitin ligase that ubiquitinates antagonists of inhibitor of apoptosis proteins SMAC, HtrA2, and ARTS. The Journal of Biological Chemistry. 2013;**288**(17):12014-12021

[34] Saul FA, Poljak RJ. Crystal structure of human immunoglobulin fragment fab new refined at 2.0 a resolution. Proteins. 1992;**14**(3):363-371

[35] Singh S, Ng J, Nayak D, Sivaraman J. Structural insights into a HECT-type E3 ligase AREL1 and its ubiquitination activities in vitro. The Journal of Biological Chemistry. 2019;**294**(52): 19934-19949

[36] Chen J, Wang B, Wu Y. Structural characterization and function prediction of immunoglobulin-like fold in cell adhesion and cell signaling. Journal of Chemical Information and Modeling. 2018;**58**(2):532-542

[37] Halaby DM, Poupon A, Mornon J. The immunoglobulin fold family: Sequence analysis and 3D structure comparisons. Protein Engineering. 1999;**12**(7):563-571

[38] Elmore S. Apoptosis: A review of programmed cell death. Toxicologic Pathology. 2007;**35**(4):495-516

[39] D'Arcy MS. Cell death: A review of the major forms of apoptosis, necrosis and autophagy. Cell Biology International. 2019;**43**(6):582-592

[40] McIlwain DR, Berger T, Mak TW. Caspase functions in cell death and disease. Cold Spring Harbor Perspectives in Biology. 2013;**5**(4): a008656

[41] Wu CC, Bratton SB. Regulation of the intrinsic apoptosis pathway by reactive oxygen species. Antioxidants & Redox Signaling. 2013;**19**(6):546-558

[42] Falschlehner C, Emmerich CH, Gerlach B, Walczak H. TRAIL signalling: Decisions between life and death. The International Journal of Biochemistry & Cell Biology. 2007;**39**(7-8):1462-1475

[43] Walczak H. Death receptor-ligand systems in cancer, cell death, and inflammation. Cold Spring Harbor Perspectives in Biology. 2013;**5**(5):a008698

[44] Walczak H, Haas TL. Biochemical analysis of the native TRAIL death-inducing signaling complex. Methods in Molecular Biology. 2008;**414**:221-239

[45] Lotan R, Rotem A, Gonen H, Finberg JP, Kemeny S, Steller H, et al. Regulation of the proapoptotic ARTS protein by ubiquitin-mediated degradation. The Journal of Biological Chemistry. 2005;**280**(27):25802-25810

[46] Michel MA, Elliott PR, Swatek KN, Simicek M, Pruneda JN, Wagstaff JL, et al. Assembly and specific recognition of k29- and k33-linked polyubiquitin. Molecular Cell. 2015;**58**(1):95-109

[47] Zhu H, Li Y, Liu Y, Han B. Bivalent SMAC mimetics for treating cancer by antagonizing inhibitor of apoptosis proteins. ChemMedChem. 2019;**14**(23):1951-1962

[48] Ali R, Singh S, Haq W. IAP proteins antagonist: An introduction and chemistry of Smac mimetics under clinical development. Current Medicinal Chemistry. 2018;**25**(31):3768-3795

[49] Fulda S. Smac mimetics to therapeutically target IAP proteins in cancer. International Review of Cell and Molecular Biology. 2017;**330**:157-169

[50] Anglesio MS, Evdokimova V, Melnyk N, Zhang L, Fernandez CV, Grundy PE, et al. Differential expression of a novel ankyrin containing E3 ubiquitin-protein ligase, Hace1, in sporadic Wilms' tumor versus normal kidney. Human Molecular Genetics. 2004;**13**(18):2061-2074

[51] El-Naggar AM, Clarkson PW, Negri GL, Turgu B, Zhang F, Anglesio MS, et al. HACE1 is a potential tumor suppressor in osteosarcoma. Cell Death & Disease. 2019;**10**(1):21

[52] Li JC, Chang X, Chen Y, Li XZ, Zhang XL, Yang SM, et al. Loss of the tumor suppressor HACE1 contributes to cancer progression. Current Drug Targets. 2019;**20**(10):1018-1028

[53] Yu Z, Li Y, Han T, Liu Z. Demethylation of the HACE1 gene promoter inhibits the proliferation of human liver cancer cells. Oncology Letters. 2019;**17**(5):4361-4368

[54] Zhou Z, Zhang HS, Zhang ZG, Sun HL, Liu HY, Gou XM, et al. Loss of HACE1 promotes colorectal cancer cell migration via upregulation of YAP1. Journal of Cellular Physiology. 2019;**234**(6):9663-9672

[55] Tikhonova EB, Ethayathulla AS, Su Y, Hariharan P, Xie S, Guan L. A transcription blocker isolated from a designed repeat protein combinatorial library by in vivo functional screen. Scientific Reports. 2015;**5**:8070

[56] Li J, Mahajan A, Tsai MD. Ankyrin repeat: A unique motif mediating protein-protein interactions. Biochemistry. 2006;**45**(51):15168-15178

[57] Tang D, Xiang Y, De Renzis S, Rink J, Zheng G, Zerial M, et al. The ubiquitin ligase HACE1 regulates Golgi membrane dynamics during the cell cycle. Nature Communications. 2011;**2**:501

[58] Lachance V, Degrandmaison J, Marois S, Robitaille M, Genier S, Nadeau S, et al. Ubiquitylation and activation of a Rab GTPase is promoted by a beta(2)AR-HACE1 complex. Journal of Cell Science. 2014;**127**(Pt 1):111-123

[59] Liu Z, Chen P, Gao H, Gu Y, Yang J, Peng H, et al. Ubiquitylation of autophagy receptor optineurin by HACE1 activates selective autophagy for tumor suppression. Cancer Cell. 2014;**26**(1):106-120

[60] Tortola L, Nitsch R, Bertrand MJM, Kogler M, Redouane Y, Kozieradzki I, et al. The tumor suppressor Hace1 is a critical regulator of TNFR1-mediated cell fate. Cell Reports. 2016;**16**(12):3414

[61] Acosta MI, Urbach S, Doye A, Ng YW, Boudeau J, Mettouchi A, et al. Group-I PAKs-mediated phosphorylation of HACE1 at serine 385 regulates its oligomerization state and Rac1 ubiquitination. Scientific Reports. 2018;**8**(1):1410

[62] Andrio E, Lotte R, Hamaoui D, Cherfils J, Doye A, Daugaard M, et al. Identification of cancer-associated missense mutations in hace1 that impair cell growth control and Rac1 ubiquitylation. Scientific Reports. 2017;**7**:44779

[63] Castillo-Lluva S, Tan CT, Daugaard M, Sorensen PH, Malliri A. The tumour suppressor HACE1 controls cell migration by regulating Rac1 degradation. Oncogene. 2013;**32**(13):1735-1742

[64] Daugaard M, Nitsch R, Razaghi B, McDonald L, Jarrar A, Torrino S, et al. Hace1 controls ROS generation of vertebrate Rac1-dependent NADPH oxidase complexes. Nature Communications. 2013;**4**:2180

[65] Torrino S, Visvikis O, Doye A, Boyer L, Stefani C, Munro P, et al. The E3 ubiquitin-ligase HACE1 catalyzes the ubiquitylation of active Rac1. Developmental Cell. 2011;**21**(5):959-965

[66] Hollstein R, Parry DA, Nalbach L, Logan CV, Strom TM, Hartill VL, et al. HACE1 deficiency causes an autosomal recessive neurodevelopmental syndrome. Journal of Medical Genetics. 2015;**52**(12):797-803

[67] Zhang L, Chen X, Sharma P, Moon M, Sheftel AD, Dawood F, et al. HACE1-dependent protein degradation provides cardiac protection in response to haemodynamic stress. Nature Communications. 2014;**5**:3430

[68] Zohn IE, Anderson KV, Niswander L. The Hectd1 ubiquitin ligase is required for development of the head mesenchyme and neural tube closure. Developmental Biology. 2007;**306**(1):208-221

[69] Madhurantakam C, Varadamsetty G, Grutter MG, Pluckthun A, Mittl PR. Structure-based optimization of designed armadillo-repeat proteins. Protein Science. 2012;**21**(7):1015-1028

[70] Zhou Z, Du X, Cai Z, Song X, Zhang H, Mizuno T, et al. Structure of Sad1-UNC84 homology (SUN) domain defines features of molecular bridge in nuclear envelope. The Journal of Biological Chemistry. 2012;**287**(8):5317-5326

[71] Escobar-Cabrera E, Lau DK, Giovinazzi S, Ishov AM, McIntosh LP. Structural characterization of the DAXX N-terminal helical bundle domain and its complex with Rassf1C. Structure. 2010;**18**(12):1642-1653

[72] Fang S, Guo H, Cheng Y, Zhou Z, Zhang W, Han B, et al. circHECTD1 promotes the silica-induced pulmonary endothelial-mesenchymal transition via HECTD1. Cell Death & Disease. 2018;**9**(3):396

[73] Thiery JP, Acloque H, Huang RY, Nieto MA. Epithelial-mesenchymal transitions in development and disease. Cell. 2009;**139**(5):871-890

[74] Peng X, Jing P, Chen J, Xu L. The role of circular RNA HECTD1 expression in disease risk, disease severity, inflammation, and recurrence of acute ischemic stroke. Journal of Clinical Laboratory Analysis. 2019;**33**(7):e22954

[75] de Caestecker MP, Yahata T, Wang D, Parks WT, Huang S, Hill CS, et al. The Smad4 activation domain (SAD) is a proline-rich, p300-dependent transcriptional activation domain. The Journal of Biological Chemistry. 2000;**275**(3):2115-2122

[76] Itoh M, Kim CH, Palardy G, Oda T, Jiang YJ, Maust D, et al. Mind bomb is a ubiquitin ligase that is essential for efficient activation of notch signaling by Delta. Developmental Cell. 2003;**4**(1):67-82

[77] D'Alonzo D, Emch FH, Shen X, Bruder E, De Geyter C, Zhang H. Hectd1 is essential for embryogenesis in mice. Gene Expression Patterns. 2019;**34**: 119064

[78] Sarkar AA, Nuwayhid SJ, Maynard T, Ghandchi F, Hill JT, Lamantia AS, et al. Hectd1 is required for development of the junctional zone of the placenta. Developmental Biology. 2014;**392**(2):368-380

[79] Sarkar AA, Zohn IE. Hectd1 regulates intracellular localization and secretion of Hsp90 to control cellular behavior of the cranial mesenchyme. The Journal of Cell Biology. 2012;**196**(6):789-800

[80] Kraja AT, Chasman DI, North KE, Reiner AP, Yanek LR, Kilpelainen TO, et al. Pleiotropic genes for metabolic syndrome and inflammation. Molecular Genetics and Metabolism. 2014;**112**(4):317-338

[81] Kittler R, Shiang C, Hutchinson R, Kollipara RK, Kapur P, Franto F, et al. Grade progression in urothelial carcinoma can occur with high or low mutational homology: A first-step toward tumor-specific care in initial low-grade bladder cancer. Oncotarget. 2018;**9**(10):9415-9424

[82] Li L, Chen Q, Fan QB, Wang S, Shi HH, Zhu L, et al. Pathogenetic gene changes of eutopic endometrium in patients with ovarian endometriosis. Chinese Medical Journal (England). 2019;**132**(9):1107-1109

[83] Wang X, Li J, Duan Y, Wu H, Xu Q, Zhang Y. Whole genome sequencing analysis of lung adenocarcinoma in Xuanwei, China. Thoracic Cancer. 2017;**8**(2):88-96

[84] Zhang H, Mo XB, Xu T, Bu XQ, Lei SF, Zhang YH. Novel genes affecting blood pressure detected via gene-based association analysis. G3 (Bethesda). 2015;**5**(6):1035-1042

[85] Brooks WS, Banerjee S, Crawford DF. G2E3 is a nucleo-cytoplasmic shuttling protein with DNA damage responsive localization. Experimental Cell Research. 2007;**313**(4):665-676

[86] Brooks WS, Helton ES, Banerjee S, Venable M, Johnson L, Schoeb TR, et al. G2E3 is a dual function ubiquitin ligase required for early

embryonic development. The Journal of Biological Chemistry. 2008;**283**(32):22304-22315

[87] Schmidt F, Karnitz LM, Dobbelstein M. G2E3 attenuating replicative stress. Aging (Albany NY). 2015;7(8):527-528

[88] Schmidt F, Kunze M, Loock AC, Dobbelstein M. Screening analysis of ubiquitin ligases reveals G2E3 as a potential target for chemosensitizing cancer cells. Oncotarget. 2015;**6**(2): 617-632

[89] Sanchez R, Zhou MM. The PHD finger: A versatile epigenome reader. Trends in Biochemical Sciences. 2011;**36**(7):364-372

[90] Chung HR, Xu C, Fuchs A, Mund A, Lange M, Staege H, et al. PHF13 is a molecular reader and transcriptional co-regulator of H3K4me2/3. eLife. 2016;5:e10607

[91] Dove KK, Stieglitz B, Duncan ED, Rittinger K, Klevit RE. Molecular insights into RBR E3 ligase ubiquitin transfer mechanisms. EMBO Reports. 2016;**17**(8):1221-1235

[92] Spratt DE, Walden H, Shaw GS. RBR E3 ubiquitin ligases: New structures, new insights, new questions. The Biochemical Journal. 2014;**458**(3):421-437

[93] Dasari S, Tchounwou PB. Cisplatin in cancer therapy: Molecular mechanisms of action. European Journal of Pharmacology. 2014;**740**:364-378

[94] Aravind L. The WWE domain: A common interaction module in protein ubiquitination and ADP ribosylation. Trends in Biochemical Sciences. 2001;**26**(5):273-275

[95] Kajiro M, Tsuchiya M, Kawabe Y, Furumai R, Iwasaki N, Hayashi Y, et al. The E3 ubiquitin ligase activity of Trip12 is essential for mouse embryogenesis. PLoS One. 2011;**6**(10):e25871

[96] Liu X, Yang X, Li Y, Zhao S, Li C, Ma P, et al. Trip12 is an E3 ubiquitin ligase for USP7/HAUSP involved in the DNA damage response. FEBS Letters. 2016;**590**(23):4213-4222

[97] Park Y, Yoon SK, Yoon JB. TRIP12 functions as an E3 ubiquitin ligase of APP-BP1. Biochemical and Biophysical Research Communications. 2008;**374**(2):294-298

[98] Christen F, Hoyer K, Yoshida K, Hou HA, Waldhueter N, Heuser M, et al. Genomic landscape and clonal evolution of acute myeloid leukemia with t(8,21): An international study on 331 patients. Blood. 2019;**133**(10): 1140-1151

[99] Hanoun N, Fritsch S, Gayet O, Gigoux V, Cordelier P, Dusetti N, et al. The E3 ubiquitin ligase thyroid hormone receptor-interacting protein 12 targets pancreas transcription factor 1a for proteasomal degradation. The Journal of Biological Chemistry. 2014;**289**(51):35593-35604

[100] Georges A, Marcon E, Greenblatt J, Frappier L. Identification and characterization of USP7 targets in cancer cells. Scientific Reports. 2018;**8**(1):15833

[101] Wang L, Zhang P, Molkentine DP, Chen C, Molkentine JM, Piao H, et al. TRIP12 as a mediator of human papillomavirus/p16-related radiation enhancement effects. Oncogene. 2017;**36**(6):820-828

[102] Bramswig NC, Ludecke HJ, Pettersson M, Albrecht B, Bernier RA, Cremer K, et al. Identification of new TRIP12 variants and detailed clinical evaluation of individuals with non-syndromic intellectual disability with or without autism. Human Genetics. 2017;**136**(2):179-192

Chapter 6

Abnormal Ubiquitination of Ubiquitin-Proteasome System in Lung Squamous Cell Carcinomas

Xianquan Zhan and Miaolong Lu

Abstract

Ubiquitination is an important post-translational modification. Abnormal ubiquitination is extensively associated with cancers. Lung squamous cell carcinoma (LUSC) is the most common pathological type of lung cancer, with unclear molecular mechanism and the poor overall prognosis of LUSC patient. To uncover the existence and potential roles of ubiquitination in LUSC, label-free quantitative ubiquitomics was performed in human LUSC vs. control tissues. In total, 627 ubiquitinated proteins (UPs) with 1209 ubiquitination sites were identified, including 1133 (93.7%) sites with quantitative information and 76 (6.3%) sites with qualitative information. KEGG pathway enrichment analysis found that UPs were significantly enriched in ubiquitin-mediated proteolysis pathway (hsa04120) and proteasome complex (hsa03050). Further analysis of 400 differentially ubiquitinated proteins (DUPs) revealed that 11 subunits of the proteasome complex were differentially ubiquitinated. These findings clearly demonstrated that ubiquitination was widely present in the ubiquitin-proteasome pathway in LUSCs. At the same time, abnormal ubiquitination might affect the function of the proteasome to promote tumorigenesis and development. This book chapter discussed the status of protein ubiquitination in the ubiquitin-proteasome system (UPS) in human LUSC tissues, which offered the scientific data to elucidate the specific molecular mechanisms of abnormal ubiquitination during canceration and the development of anti-tumor drugs targeting UPS.

Keywords: lung squamous cell carcinoma, ubiquitination, ubiquitinated protein (UP), differentially ubiquitinated protein (DUP), ubiquitin-proteasome system (UPS)

1. Introduction

Ubiquitination is one of the important protein post-translational modifications (PTMs) in human body, in which ubiquitin, a 76-amino-acid protein with a molecular weight of 8.5 KDa, is covalently attached its C-terminus to the ε-amino group of the substrate protein lysine residue through a multi-step enzymatic reaction cascade catalyzed by ubiquitin-activating enzymes (E1), ubiquitin-conjugating enzymes (E2), and ubiquitin ligases (E3) [1]. As substrate proteins commonly contain multiple lysine residues, there are a variety of ubiquitination forms such as monoubiquitination (only one ubiquitin attached to a protein), multiubiquitination (several lysine residues of substrate proteins were tagged with single ubiquitin), and polyubiquitination (a polyubiquitin chain is derived from subsequent ubiquitin covalently attached to lysine residues or N-terminus of the former ubiquitin) [2]. It is worth noting that the ubiquitin itself also has seven lysine residues to greatly complicate the topology of the polyubiquitin chain. Different ubiquitination forms perform different functions, such as monoubiquitination or multiubiquitination has been shown to be required for the entry of certain cargo proteins into vesicles at different stages of the secretory/endocytic pathway, while lysine-48 ubiquitin chain is mainly related to proteasome [3]. Like other PTMs, ubiquitination is a reversible reaction, and there are over 100 deubiquitination enzymes that regulate this process [4]. Ubiquitination coordinates with deubiquitination to regulate a broad host of cellular processes, including DNA repair, cell differentiation, signal transduction, enzymatic activity regulation, assembly of multiprotein complexes, protein trafficking, and autophagy [5]. Therefore, abnormal ubiquitination is associated with many diseases, including cancer, neurodegenerative disease, infection, and immune disorders [6]. Considering the importance of ubiquitination in tumorigenesis, different components of ubiquitin-proteasome system could be regarded as targets for discovery of anti-tumor drugs. With the application of first and second therapeutic proteasome inhibitors, such as Bortezomib (FDA has approved it for multiple myeloma and mantle cell lymphoma) [7] and Carfilzomib (FDA has approved it for relapsed and refractory multiple myeloma) [8], more and more anti-tumor drugs targeting UPS have been developed and approved by FDA, such as thalidomide, lenalidomide, and pomalidomide for treatment of multiple myeloma [9, 10].

Lung squamous cell carcinoma (LUSC) is a common type of lung cancer without a clear molecular mechanism. Currently, surgery, radiation, and chemotherapy have made significant advances in lung cancer treatment, especially targeted drug therapy; for example, epidermal growth factor receptor (EGFR) mutation or EML4-ALK fusion-based targeted therapies have improved the survival time of patients with lung adenocarcinoma (LUAD). However, targeted therapy and early-stage diagnosis are still a big clinical challenge in LUSC patients [11]. Although FGFR1 amplification and DDR2 mutation have been nominated as "druggable" targets in LUSC patients, the clinical efficacies of the corresponding drugs are still under clinical trials [12, 13]. Considering that abnormal ubiquitination will lead to the occurrence of a variety of tumors and the widespread clinical applications of anti-tumor drugs for the ubiquitin-proteasome pathway in recent years, the study of quantitative ubiquitinomics in LUSC tissues may provide the direction for the development of biomarkers and new targeted drugs.

High-resolution liquid chromatography in combination with tandem mass spectrometry (LC-MS/MS) has been used as a power tool for large-scale identification of various PTMs such as ubiquitination, phosphorylation, acetylation, and N-glycosylation [14]. The challenge of the use of LC-MS/MS to identify endogenous ubiquitination sites on a large scale is the fact that ubiquitination is a low abundance event *in vivo* and the size of the modification itself. The common strategy to identify low-abundance ubiquitinations in a proteome is that the extracted protein sample is firstly digested with trypsin to form tryptic peptide mixture, then the commercially specific anti-K-ε-GG antibodies are used to preferentially enrich ubiquitinated peptides from tryptic peptide mixture before MS/MS analysis in recent years [15]. Anti-ubiquitin antibody (specific anti-K-ε-GG group)-based label-free quantification coupled with LC-MS/MS has been used as an effective method to detect, identify, and quantify ubiquitinated proteins and ubiquitination sites, and more than 10,000 ubiquitination sites have been identified and quantified [16]. For lung cancer, ubiquitinomics is mainly carried out in lung cancer cells [17, 18], while the ubiquitinomics of fresh LUSC tissues is only reported recently in our research group with label-free quantitative proteomics method and bioinformatics analysis to reveal the functions of ubiquitinome in predictive, preventive, and personalized medicine (PPPM) of LUSC [19].

This book chapter mainly reviewed ubiquitinated proteins (UPs) and differentially ubiquitinated proteins (DUPs) in ubiquitin-proteasome-system (UPS) in LUSC, and emphasized the potential regulatory role of ubiquitination in UPS, which offers scientific data for further research on the regulatory mechanism of ubiquitination on UPS, the molecular mechanism of UPS abnormality in tumor development, and the development of anti-tumor drugs targeting UPS.

2. Materials and methods

2.1 Lung cancer tissues and protein extraction

Human LUSC tissues (n = 5) and tumor-adjacent control lung tissues (n = 5) were surgically removed from patients, immediately stored in liquid nitrogen (−196°C), and then stored in freezer (−80°C). Clinical characteristics of each sample were described previously [19]. LUSC tissues (750 mg, equally mixed 5 tumor tissues) and control tissues (750 mg, equally mixed 5 control tissues) were washed 5 times with 3 mL 0.9% NaCl to clean blood on the surface of the tissues. The washed tissues (LUSC; or controls) were homogenized with urea lysis buffer [2 M thiourea, 7 M urea, 1 mM protein inhibitor PMSF, and 100 mM dithiothreitol(DTT)], sonicated, and centrifuged (15,000 g, 20 min, and 4°C). The supernatant was the extracted protein sample. The protein content was tested with Brandford method. The detailed procedure of protein extraction was described previously [19].

2.2 Protein digestion and enrichment of ubiquitinated peptides

An amount of DTT (final concentration = 10 mM) was added to each extracted protein sample, which was mixed (600 rpm, 1.5 h, and room temperature). An amount of iodoacetamide (final concentration = 50 mM) was added to the DTT-treated protein sample, which was incubated (dark, 30 min). The uranyl acetate (UA) was diluted to 2 M with 50 mM Tris HCl buffer (pH 8.0) and added to each protein sample. An amount of trypsin was added to each protein sample (trypsin:protein = 1:50 at wt:wt), and then incubated (37°C, 15–18 h). A volume of 10% trifluoroacetic acid (TFA; final concentration = 0.1%) was added, and pH was adjusted to ≤ 3 to stop digestion. Each tryptic peptide sample was purified with C18 cartridges and then lyophilized. The lyophilized tryptic peptides were resolved with 1.4 mL immunoaffinity purification (IAP) buffer that contained 50 mM NaCl, 10 mM Na_2HPO_4, 50 mM MOPS/NaOH, and pH 7.2. The anti-K-ε-GG antibodies against ubiquitin remnant motif (K-ε-GG) (Cell Signal Technology) were used to enrich the ubiquitinated peptides, followed by purification with C18 STAGE Tips. The purified ubiquitinated peptide sample was used for MS/MS analysis. The detailed experimental procedure was described previously [19].

2.3 LC-MS/MS

The prepared ubiquitinated peptide sample was analyzed with LC-MS/MS in the Easy nLC and Q Exactive mass spectrometer (Thermo Scientific). The MS/MS data for each sample were used to search protein database using MaxQuant 1.5.3.17 software to identify ubiquitinated proteins and ubiquitination sites and quantify the abundance of ubiquitination. The detailed procedure was described previously [19].

2.4 Bioinformatics

For UPs and DUPs, DAVID software (version 6.8, https://david.ncifcrf.gov/) was used to carry out the gene ontology (GO) enrichment analysis, including cellular components (CCs), molecular functions (MFs), and biological processes (BPs), and group those proteins into different functional clusters [20]. The statistically significant pathways were mined with the Kyoto Encyclopedia of Genes and Genomes (KEGG) pathway enrichment analysis. The KEGG online service tool KOBAS (http://kobas.cbi.pku.cn) was used to annotate the KEGG database description of each protein [21].

3. Results and discussion

3.1 Proteomics analysis of lysine-ubiquitinated profile in LUSC

To identify protein lysine-ubiquitinated sites and quantify the level of ubiquitination in human LUSC tissues, proteins were extracted and digested into peptides with trypsin. Lysine-ubiquitinated peptides were immunoaffinity-enriched with commercially specific anti-K-ε-GG antibodies and analyzed with high-resolution LC-MS/MS. In total, 1209 lysine-ubiquitinated sites in 627 unique proteins were identified. These proteins containing ubiquitinated lysine residues were defined as UPs. **Figure 1** showed two representative MS/MS spectra of the ubiquitinated peptides ^{425}ETNLDSLPLVDTHSK*R^{440} from vimentin (P08670; K* = ubiquitinated lysine residue) (**Figure 1A**), and ^{633}RPVK*DGGGTNSITVR^{647} from multidrug resistance-associated protein 1 (P33527; K* = ubiquitinated lysine residue) (**Figure 1B**). All other ubiquitinated sites and ubiquitinated proteins were identified with the same MS/MS method. The differentially ubiquitinated peptides were determined with amino acid sequences, ratio(tumor/control) > 2.0 or < 0.5, and p-value < 0.05. Proteins containing this type of ubiquitinated peptides were defined as DUPs. Totally, 400 DUPs with 654 ubiquitinated sites were identified in LUSC tissues vs. tumor-adjacent control lung tissues [19].

3.2 UPs and DUPs were significantly enriched in UPS-related biological processes and molecular functions in LUSC

GO functional enrichment analyses of 627 UPs and 400 DUPs were carried out according to BPs, MFs, and CCs. GO enrichment result-based cluster analysis grouped those UPs into seven clusters (**Table 1**), and DUPs into 10 clusters (**Table 2**).

Among GO enrichment results of 627 UPs, many biological processes, molecular functions, and cellular components related to UPS were significantly enriched, including negative regulation of ubiquitin-protein ligase activity involved in mitotic cell cycle, and positive regulation of ubiquitin-protein ligase activity involved in regulation of mitotic cell cycle transition in cluster 2, and proteasome-activating ATPase activity, positive regulation of proteasomal protein catabolic process, cytosolic proteasome complex, nuclear proteasome complex, and proteasome regulatory base complex in cluster 3 (**Table 1**). Interestingly, DUPs were also significantly enriched in the similar biological processes, molecular functions, and cellular components, including proteasome accessory complex, negative regulation of ubiquitin-protein ligase activity involved in mitotic cell cycle, positive regulation of ubiquitin-protein ligase activity involved in regulation of mitotic cell cycle transition, and protein polyubiquitination in cluster 2, and proteasome-activating ATPase activity, positive regulation of proteasomal protein catabolic process, cytosolic proteasome complex, nuclear proteasome complex, and proteasome regulatory

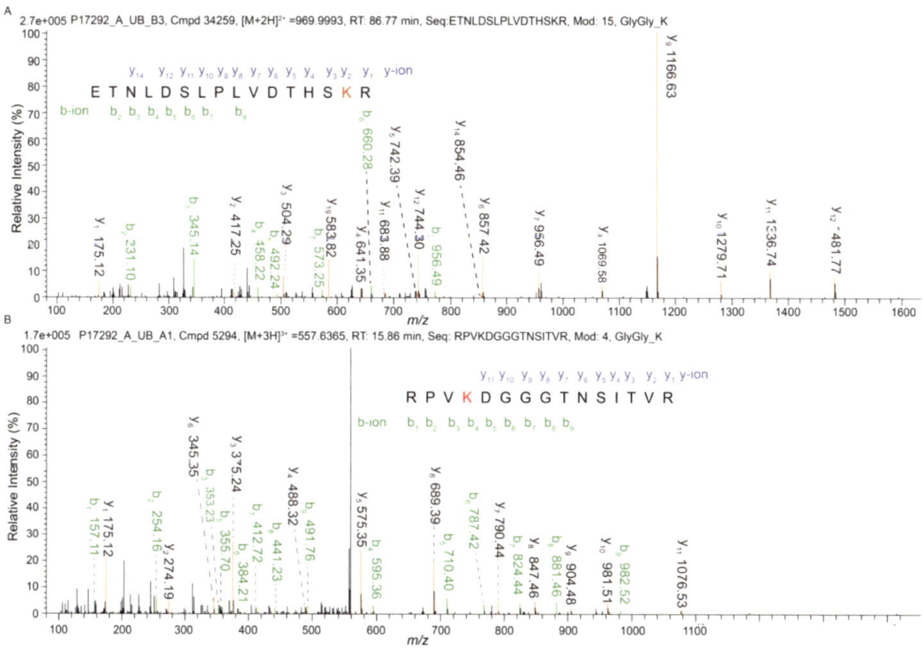

Figure 1.
*Representative MS/MS spectra of ubiquitinated peptides ^{425}ETNLDSLPLVDTHSK*R^{440} from vimentin (P08670) (A) and ^{633}RPVK*DGGGTNSITVR^{647} from multidrug resistance-associated protein 1 (MRP1) (P33527) (B). K* = ubiquitinated lysine residue. Reproduced from Lu et al. [19], with permission from Springer publisher open access article, copyright 2020.*

base complex in cluster 3 (**Table 2**). These findings clearly demonstrated that many ubiquitinated proteins were involved in UPS system, and differential ubiquitination occurred in UPS system in LUSC, implying that ubiquitination participated in the regulation of UPS, and abnormal ubiquitination might play an important role in the development of LUSC.

3.3 UPs involved in UPS-related molecular network alternations in LUSC

KEGG pathway network analysis of 627 UPs revealed 47 statistically significant ubiquitination-mediated signaling pathway alterations ($P < 0.05$ and $FDR < 0.05$) (**Figure 2**), among which were included two UPS-related pathways—ubiquitin-mediated proteolysis pathway (hsa04120) and proteasome complex (hsa03050).

Ubiquitin-mediated proteolysis pathway showed the detailed process of protein ubiquitination, which involved multiple types of E1s, E2s, and E3s. This study found that one E1 (UBE1), two E2s (UBE2N, and UBE2O), and six E3s (ITCH, HUWE1, UBE4B, PML, CUL4A, and CUL5) were ubiquitinated in LUSC (**Figure 3**). These six E3s belonged to different subfamilies, in which ITCH and HUWE1 belonged to HECT type E3, UBE4B belonged to U-box type E3, PML belonged to single RING-finger type E3, and both CUL4A and CUL5 belonged to multi subunit RING-finger type E3. E1s, E2s, and E3s are the important enzymes to catalyze the occurrence of ubiquitination in a protein. The ubiquitination of these enzymes definitely affects the quitination process of a protein. Currently, studies on these enzymes have focused on their roles in the ubiquitination process, and the effects of PTMs on these enzymes are poorly understood. There are relatively few studies on the ubiquitination of these nine enzymes. For example, ubiquitinated PML (P29590, identified in this study) was mediated by multiple E3s, leading to subsequent proteasomal degradation [22, 23]. Self-ubiquitination of ITCH (Q96J02, identified in this study)

Category	GO term	p -value
Annotation cluster 1		
GOTERM_MF_DIRECT	Cadherin binding involved in cell-cell adhesion	1.43E-29
GOTERM_CC_DIRECT	Cell-cell adherens junction	3.72E-29
GOTERM_BP_DIRECT	Cell-cell adhesion	1.13E-23
Annotation cluster 2		
GOTERM_BP_DIRECT	Wnt signaling pathway, planar cell polarity pathway	5.78E-11
GOTERM_BP_DIRECT	Regulation of cellular amino acid metabolic process	4.14E-09
GOTERM_BP_DIRECT	Stimulatory C-type lectin receptor signaling pathway	4.43E-09
GOTERM_BP_DIRECT	NIK/NF-kappa B signaling	1.37E-08
GOTERM_BP_DIRECT	Tumor necrosis factor-mediated signaling pathway	2.96E-08
GOTERM_BP_DIRECT	Negative regulation of ubiquitin-protein ligase activity involved in mitotic cell cycle	3.70E-08
GOTERM_BP_DIRECT	Positive regulation of ubiquitin-protein ligase activity involved in regulation of mitotic cell cycle transition	9.13E-08
GOTERM_BP_DIRECT	Anaphase-promoting complex-dependent catabolic process	1.52E-07
GOTERM_BP_DIRECT	T cell receptor signaling pathway	9.74E-07
GOTERM_BP_DIRECT	Positive regulation of canonical Wnt signaling pathway	1.13E-06
GOTERM_BP_DIRECT	Negative regulation of canonical Wnt signaling pathway	1.57E-05
Annotation cluster 3		
GOTERM_CC_DIRECT	Proteasome regulatory particle, base subcomplex	1.29E-08
GOTERM_CC_DIRECT	Nuclear proteasome complex	1.70E-08
GOTERM_CC_DIRECT	Cytosolic proteasome complex	1.21E-07
GOTERM_MF_DIRECT	Proteasome-activating ATPase activity	1.70E-07
GOTERM_BP_DIRECT	Positive regulation of RNA polymerase II transcriptional preinitiation complex assembly	3.45E-06
GOTERM_MF_DIRECT	TBP-class protein binding	3.27E-05
GOTERM_BP_DIRECT	Positive regulation of proteasomal protein catabolic process	9.97E-05
Annotation cluster 4		
GOTERM_BP_DIRECT	SRP-dependent cotranslational protein targeting to membrane	3.56E-08
GOTERM_BP_DIRECT	Viral transcription	7.74E-08
GOTERM_BP_DIRECT	Nuclear-transcribed mRNA catabolic process, nonsense-mediated decay	1.92E-07
GOTERM_BP_DIRECT	Translational initiation	6.60E-06
GOTERM_CC_DIRECT	Ribosome	3.32E-05
GOTERM_MF_DIRECT	Structural constituent of ribosome	1.68E-03
Annotation cluster 5		
GOTERM_CC_DIRECT	Haptoglobin-hemoglobin complex	9.95E-05
Annotation cluster 6		
GOTERM_BP_DIRECT	Nucleotide-excision repair, DNA damage recognition	6.40E-04
Annotation cluster 7		
GOTERM_CC_DIRECT	Hemoglobin complex	4.59E-03

Table 1.
The functional categories of 627 ups, identified with GO enrichment analysis.

Category	GO term	p-value
Annotation cluster 1		
GOTERM_MF_DIRECT	Cadherin binding involved in cell-cell adhesion	3.11E-18
GOTERM_CC_DIRECT	Cell-cell adherens junction	2.24E-17
GOTERM_BP_DIRECT	Cell-cell adhesion	1.23E-13
Annotation cluster 2		
GOTERM_CC_DIRECT	Proteasome accessory complex	4.60E-12
GOTERM_BP_DIRECT	Antigen processing and presentation of exogenous peptide antigen via MHC class I, TAP-dependent	1.59E-11
GOTERM_BP_DIRECT	Regulation of cellular amino acid metabolic process	3.49E-09
GOTERM_BP_DIRECT	NIK/NF-kappaB signaling	5.35E-09
GOTERM_BP_DIRECT	Negative regulation of ubiquitin-protein ligase activity involved in mitotic cell cycle	1.27E-08
GOTERM_BP_DIRECT	Stimulatory C-type lectin receptor signaling pathway	1.85E-08
GOTERM_BP_DIRECT	Positive regulation of ubiquitin-protein ligase activity involved in regulation of mitotic cell cycle transition	2.83E-08
GOTERM_BP_DIRECT	Anaphase-promoting complex-dependent catabolic process	4.43E-08
GOTERM_BP_DIRECT	Tumor necrosis factor-mediated signaling pathway	8.37E-08
GOTERM_BP_DIRECT	Positive regulation of canonical Wnt signaling pathway	1.04E-07
GOTERM_BP_DIRECT	T cell receptor signaling pathway	2.37E-07
GOTERM_BP_DIRECT	Fc-epsilon receptor signaling pathway	4.90E-07
GOTERM_BP_DIRECT	Protein polyubiquitination	7.65E-07
GOTERM_BP_DIRECT	Negative regulation of canonical Wnt signaling pathway	4.36E-06
Annotation cluster 3		
GOTERM_CC_DIRECT	Proteasome regulatory particle, base subcomplex	4.88E-10
GOTERM_CC_DIRECT	Nuclear proteasome complex	1.00E-09
GOTERM_CC_DIRECT	Cytosolic proteasome complex	7.28E-09
GOTERM_MF_DIRECT	Proteasome-activating ATPase activity	1.54E-08
GOTERM_BP_DIRECT	Positive regulation of RNA polymerase II transcriptional preinitiation complex assembly	3.41E-07
GOTERM_MF_DIRECT	TBP-class protein binding	2.10E-06
GOTERM_BP_DIRECT	Positive regulation of proteasomal protein catabolic process	1.05E-05
GOTERM_BP_DIRECT	Protein catabolic process	1.41E-04
Annotation cluster 4		
GOTERM_BP_DIRECT	Regulation of ventricular cardiac muscle cell action potential	1.11E-03
GOTERM_MF_DIRECT	Cell adhesive protein binding involved in bundle of His cell-Purkinje myocyte communication	5.30E-03
Annotation cluster 5		
GOTERM_CC_DIRECT	Haptoglobin-hemoglobin complex	2.43E-05
GOTERM_CC_DIRECT	Endocytic vesicle lumen	1.71E-04
GOTERM_MF_DIRECT	Haptoglobin binding	1.10E-03
GOTERM_BP_DIRECT	Positive regulation of cell death	1.90E-02

Category	GO term	p-value
GOTERM_CC_DIRECT	Hemoglobin complex	1.97E-02
GOTERM_MF_DIRECT	Oxygen transporter activity	2.90E-02
GOTERM_BP_DIRECT	Oxygen transport	3.43E-02
Annotation cluster 6		
GOTERM_BP_DIRECT	SRP-dependent cotranslational protein targeting to membrane	9.85E-05
GOTERM_BP_DIRECT	Nuclear-transcribed mRNA catabolic process, nonsense-mediated decay	5.83E-04
GOTERM_BP_DIRECT	Viral transcription	1.69E-03
GOTERM_BP_DIRECT	Translational initiation	5.82E-03
GOTERM_CC_DIRECT	Ribosome	3.40E-02
Annotation cluster 7		
GOTERM_MF_DIRECT	Voltage-gated anion channel activity	5.30E-03
GOTERM_MF_DIRECT	Porin activity	5.30E-03
GOTERM_CC_DIRECT	Pore complex	1.11E-02
GOTERM_BP_DIRECT	Anion transport	3.02E-02
GOTERM_BP_DIRECT	Regulation of anion transmembrane transport	4.34E-02
Annotation cluster 8		
GOTERM_BP_DIRECT	Daunorubicin metabolic process	1.00E-02
GOTERM_BP_DIRECT	Doxorubicin metabolic process	1.00E-02
Annotation cluster 9		
GOTERM_BP_DIRECT	Nucleotide-excision repair, DNA damage recognition	1.00E-02
GOTERM_BP_DIRECT	Global genome nucleotide-excision repair	2.47E-02
Annotation cluster 10		
GOTERM_MF_DIRECT	Neutral amino acid transmembrane transporter activity	1.82E-02
GOTERM_BP_DIRECT	Neutral amino acid transport	2.62E-02

Table 2.
The functional categories of 400 DUPs, identified with GO enrichment analysis. Modified from Lu et al. [19], with permission from Springer publisher open access article, copyright 2020.

through lysine-63 linkages showed an auto-regulatory mechanism controlling ITCH cytoplasmic-nuclear shuffling [24]. Therefore, the effects of the currently known ubiquitination on these enzymes are only the tip of the iceberg. However, one should also realize that this study found ubiquitination of cullin proteins such as Cul4A and Cul5, while cullin proteins can also be modified by NEDD8 to form NEDDylation. It is well known that the use of the K-ε-GG antibody cannot discriminate between proteins modified with ubiquitin and the related proteins NEDD8 and ISG15. Therefore, for deep investigation of this identified ubiquitination of E3s Cul4A and Cul5 in LUSC in the future, additional experiments are needed to discriminate E3s Cul4A and Cul5 that were modified by ubiquitin, NEDD8, or ISG15 [22].

Proteasome was a pivotal component for ubiquitin-mediated proteolysis. The 26S proteasome was a complex including two 19S regulatory particles (PA700) and one 20S core particle. The 20S degradation complex contained two α rings (7 subunits, α1-α7) and two β rings (7 subunits, β1-β7). These α rings and β rings together formed a hollow ground circle. The tube-like structure was highly conserved

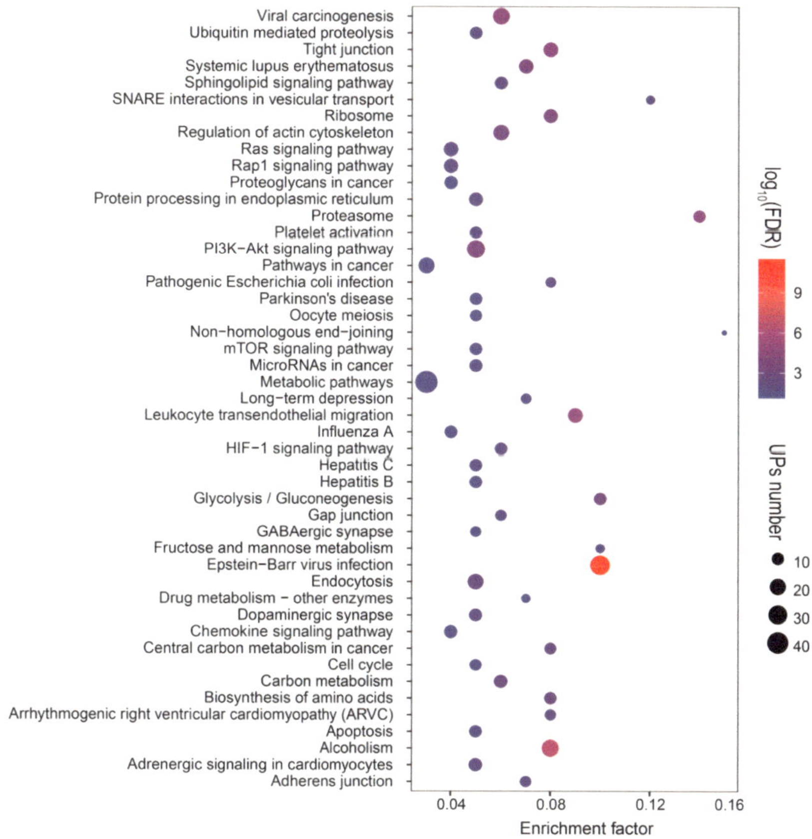

Figure 2.
Statistically significant KEGG pathways enriched from 627 UPs. Those UPs were significantly enriched in 47 KEGG pathways ($p < 0.05$ and $FDR < 0.05$), among which were included two UPS-related pathways—ubiquitin-mediated proteolysis pathway (hsa04120) and proteasome complex (hsa03050). The darker dot means the more significant enrichment, and the size of the dot represents the number of UPs enriched in the pathway.

from archaea to mammals [25]. Among them, the α ring was located in the outer layer of the cylinder-like structure, which mainly acted on the recognition of the substrate; the β ring was located in the inner layer of the cylinder-like structure, and was mainly responsible for catalyzing the degradation of the substrate [26, 27]. Three subunits that played a catalytic role were located on the inner surface of the β-ring molecule, exhibiting cysteine protease-like activity, trypsin-like activity, and chymotrypsin-like activity [26, 27]. The 19S regulatory complex contained 19 different subunits, which were divided into two parts: "base" and "lid" [28]. Among them, the base part formed the proximal part of the 19S regulatory complex, which was connected to the alpha ring of the 20S degradation complex, and the lip part formed the distal end. The base section contained 6 ATPase-dependent subunits (Rpt1-Rpt6) and 2 ATPase-independent subunits (Rpn1 and Rpn2) [28]. Usually, Rpn10 and Rpn13 were also classified as the base [26]. The lid part consisted of Rpn3, Rpn5-Rpn9, and Rpn11-Rpn12 subunits [28]. The 19S regulatory complex recognized ubiquitin-labeled target proteins (Rpn10 and Rpn13) and before the target protein entered the 20S degradation complex, deubiquitinated the target protein (Rpn11) and opened the folded structure of the target protein [28]. This study discovered five UPs (Rpn3, Rpn5, Rpn6, Rpn10, and Rpn12) in PA700 (Lid), and seven UPs (Rpn13, Rpt1, Rpt2, Rpt3, Rpt4, Rpt5, and Rpt6) in PA700 (Base) in LUSC. No UPs were identified in 20S core particle in LUSC (**Figure 4**).

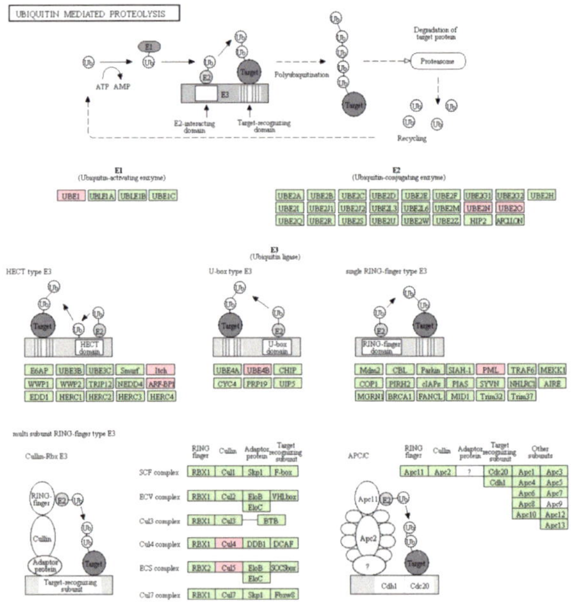

Pathway nodes	Protein access number	Modified sites	Ratio (T/N)
UBE1	A0A024RDB0	K544	2.90
UBE2N	P61088	K94	0.09
UBE2N	P61088	K92	0.26
UBE2N	P61088	K82	0.55
UBE2N	P61088	K74	NA*
UBE2O	Q9C0C9	K136	0.81
UBE2O	Q9C0C9	K132	0.93
Itch	Q96J02	K446	1.81
Itch	Q96J02	K478	1.83
Itch	Q96J02	K432	3.05
Itch	Q96J02	K350	NA*
ARF-BP1	A0A024R9W5	K3452	2.28
ARF-BP1	A0A024R9W5	K4295	NA*
UBE4B	O95155	K794	NA*
PML	P29590	K394	1.33
Cul4	Q13619	K265	NA*
Cul4	Q13619	K465	NA*
Cul4	Q13619	K635	NA*
Cul5	Q93034	K724	0.96

Figure 3.
UPs identified in ubiquitin-mediated proteolysis pathway (hsa04120). Pink rectangle means the ubiquitinated proteins, and green rectangle means nonubiquitinated proteins. The pathway node in the right panel corresponds to the pink rectangle in the left diagram. Protein access number is the Swiss-Prot accession number. Modified sites refer to ubiquitinated lysine (K). Ratio (T/N) = ratio of tumors to controls. Asterisk (*) represents there is no quantitative intensity on this modification site in both the tumor and the control group.

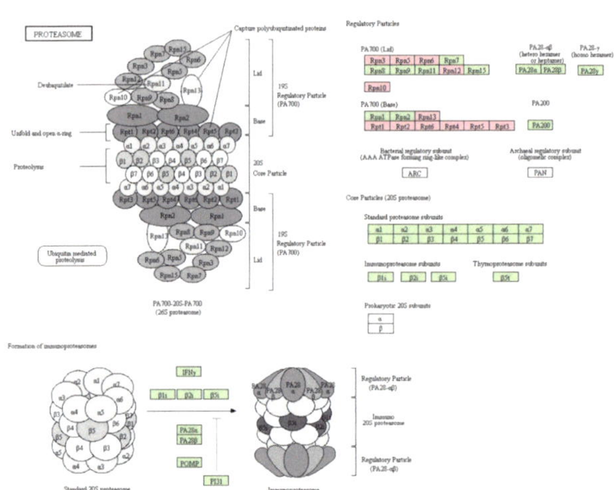

Pathway nodes	Protein access number	Modified sites	Ratio (T/N)
Rpn13	Q16186	K34	2.27
Rpt2	P62191	K293	T+/N-
Rpt1	A0A140VK70	K64	4.96
Rpt1	A0A140VK70	K415	6.57
Rpt5	A0A140VK42	K53	0.22
Rpt5	A0A140VK42	K276	4.90
Rpt5	A0A140VK42	K79	0.99
Rpt5	A0A140VK42	K372	T+/N-
Rpt3	A8K2M0	K271	T+/N-
Rpt6	A0A140VJS3	K346	T+/N-
Rpt6	A0A140VJS3	K290	T+/N-
Rpt6	A0A140VJS3	K330	T+/N-
Rpt4	A0A087X2I1	K62	2.37
Rpt4	A0A087X2I1	K194	T+/N-
Rpt4	A0A087X2I1	K328	T+/N-
Rpt4	A0A087X2I1	K211	NA*
Rpn3	O43242	K273	T+/N-
Rpn12	P48556	K111	NA*
Rpn6	O00231	K32	T+/N-
Rpn5	A0A0S2Z489	K147	T+/N-
Rpn10	P55036	K40	33.80
Rpn10	P55036	K74	5.16

Figure 4.
UPs identified in proteasome complex (hsa03050). Pink rectangle means the ubiquitinated subunits, and green rectangle means non-ubiquitinated subunits. The pathway node in the right panel corresponds to the pink rectangle in the left diagram. Protein access number is the Swiss-Prot accession number. Modified sites refer to ubiquitinated lysine (K). Ratio (T/N) = ratio of tumors to controls. T+/N– indicates that this modification site has quantitative intensity only in tumor group, while T–/N+ means this modification site has quantitative intensity only in control group. Asterisk (*) represents there is no quantitative intensity on this modification site in both the tumor and the control group.

3.4 DUPs involved in UPS-related molecular network alternations in LUSC

KEGG pathway network analysis of 400 DUPs revealed 39 statistically significant ubiquitination-mediated signaling pathway alterations ($P < 0.05$ and $FDR < 0.05$) (**Figure 5**), including one UPS-related pathway – proteasome complex (hsa03050).

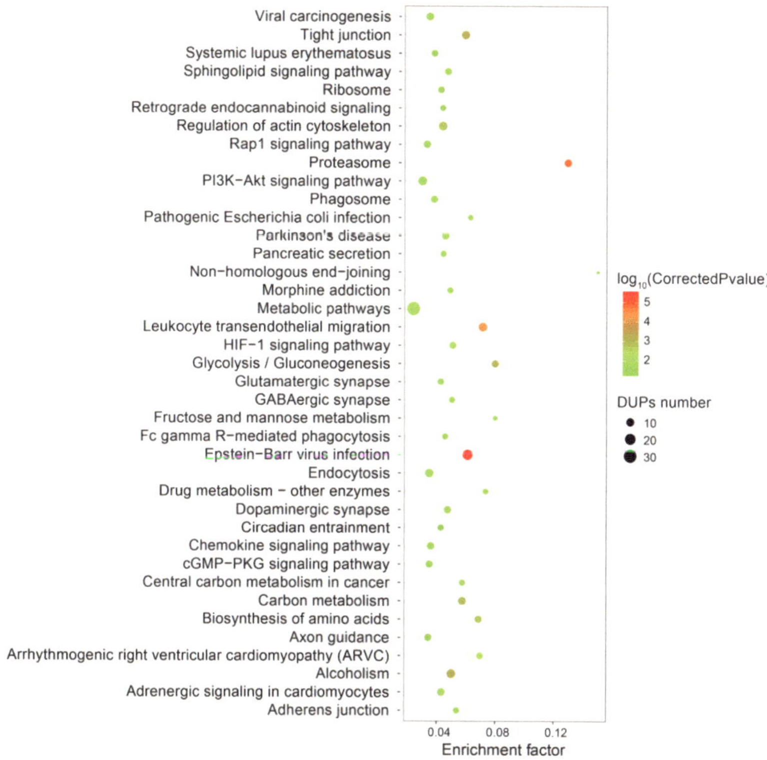

Figure 5.
Statistically significant KEGG pathways enriched from 400 DUPs. Those DUPs were significantly enriched in 39 KEGG pathways ($p < 0.05$ and FDR < 0.05), including one UPS-related pathway – proteasome complex. The darker dot means the more significant enrichment, and the size of the dot represents the number of DUPs enriched in the pathway. Reproduced from Lu et al. [19], with permission from Springer publisher open access article, copyright 2020.

In proteasome complex, this study discovered 4 DUPs (Rpn3, Rpn5, Rpn10, and Rpn6) in PA700 (Lid), and 7 DUPs (Rpn13, and Rpt1-Rpt6) in PA700 (Base). Their ubiquitination levels were significantly increased at residues K_{74} (Ratio = 5.16) in Rpn10, K_{34} (Ratio = 2.27) in Rpn13, K_{293} (T+/N−) in Rpt2, K_{46} (Ratio = 4.96) in Rpt1, K_{372} (T+/N−) in Rpt5, K_{273} (T+/N−) in Rpt4, K_{346} (T+/N−), K_{330} (T+/N−) and K_{290} (T+/N−) in Rpt6, K_{194} (T+/N−), K_{328} (T+/N−) and K_{62} (Ratio = 2.37) in Rpt4, K_{273} (T+/N−) in Rpn3, K_{32} (T+/N−) in Rpn6, and K_{147} (T+/N−) in Rpn5. The ubiquitination level was decreased at residue K_{53} (Ratio = 0.33) in Rpt5 (**Figure 6**).

The proteasome was a pivotal component of UPS to degrade the short-lived regulatory proteins and remove the damaged soluble proteins [29]. Consequently, dysfunction of proteasome might decrease the capability of protein degradation, thus resulting in the increased level of misfolded and damaged proteins, which was closely related to tumorigenesis [30]. The 26S proteasome had one 20S subunit and two 19S regulatory caps. Two 19S caps were necessary to maintain the normal functions of 20S subunit. For example, Rpn 13 in 19S base cap and Rpn 10 in 19S head cap were the recognition-receptors of the ubiquitinated proteins [31, 32]. Further, PTMs (such as phosphorylation, acetylation, myristoylation, and ubiquitination) had been detected in those subunits to greatly complicate the mechanisms of the modulation of proteasome activity. For example, Rpn 10 was mono-ubiquitinated to recruit substrate protein and interact with the shuttle factor of proteasome in drosophila [33, 34]. The multiple ubiquitinations in 19S cap of proteasome such as Rpn 1, Rpn 10, and Rpn 13 were necessary to autophage proteasome [35]. Our study [19] discovered three non-ATPase subunits

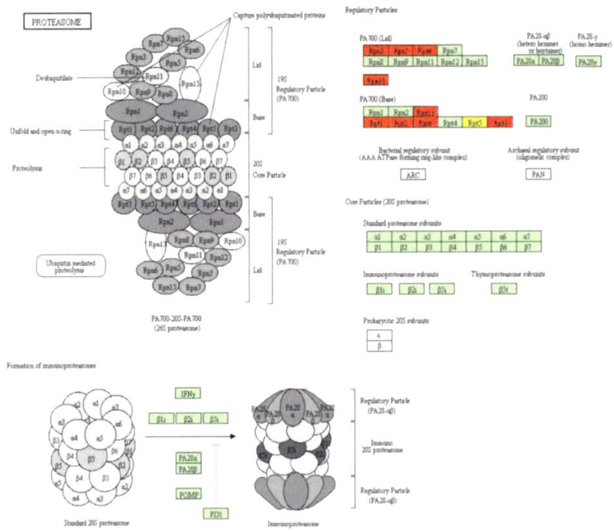

Pathway nodes	Protein access number	Modified site	Ratio (T/N)
Rpn13	Q16186	K34	2.27
Rpt2	P62191	K293	T+/N-
Rpt1	A0A140VK70	K46	4.96
Rpt5	A0A140VK42	K372	T+/N-
Rpt5	A0A140VK42	K53	0.33
Rpt4	A8K2M0	K273	T+/N-
Rpt6	A0A140VJS3	K346	T+/N-
Rpt6	A0A140VJS3	K330	T+/N-
Rpt6	A0A140VJS3	K290	T+/N-
Rpt4	A0A087X2I1	K194	T+/N-
Rpt4	A0A087X2I1	K328	T+/N-
Rpt4	A0A087X2I1	K62	2.37
Rpn3	O43242	K273	T+/N-
Rpn6	O00231	K32	T+/N-
Rpn5	A0A0S2Z489	K147	T+/N-
Rpn10	Q5VWC4	K74	5.16

Figure 6.
DUPs identified in proteasome complex (hsa03050). The red rectangle means the intensities of all identified ubiquitination sites in one protein were increased, and yellow rectangle means at least two ubiquitination sites in a protein with inconsistent ubiquitination intensities. The pathway node in the right panel corresponds to the red and yellow rectangle in the left diagram. Protein access number is the Swiss-Prot accession number. Modified sites refer to ubiquitinated lysine (K). Ratio (T/N) = ratio of tumors to controls. T+/N– indicates that this modification site has quantitative intensity only in tumor group, while T–/N+ means this modification site has quantitative intensity only in control group. Reproduced from Lu et al. [19], with permission from Springer publisher open access article, copyright 2020.

(PSMD3, PSMD11, and PSMD12), and three ATPase subunits (PSMC1, PSMC4, and PSMC6) were differentially ubiquitinated in 19S regulatory cap of proteasome in LUSC tissues. It clearly demonstrated that these ubiquitinations in 19S regulatory caps might influence the structure and functions of the proteasome complex. Some studies found that PSMD11 was necessary to assemble proteasome complex and elevate the activity of proteasome in embryonic stem cells [36]. Acetylation [37], phosphorylation [38], and SUMO [39] had been reported to occur in PSMD11, and our study first discovered that PSMD11 was ubiquitinated at residue K_{32} in LUSC tissues but not in control lung tissues [19]. Currently, few literature studies are found regarding the study on the relationship of ubiquitination and function of proteasome subunits. However, the abnormal ubiquitination of proteasome subunits might cause the functional abnormalities of proteasome complex in LUSC tissues and further lead to the imbalance of synthesis and degradation of intracellular proteins. These findings offer the new clues to deeply study and understand the regulation of UPS functions in LUSC.

4. Conclusions

Label-free quantitative ubiquitinomics was an effective approach to identify ubiquitinated proteins and ubiquitination sites and quantifies the levels of ubiquitination in human LUSC tissues. In total, 627 UPs and 400 DUPs were identified, providing the first (differential) ubiquitinome profile based on fresh human LUSC tissues. GO and KEGG analyses of UPs and DUPs revealed the statistically significant ubiquitination-mediated molecular network alternations, among which several proteins in two UPS-related pathways (ubiquitin-mediated proteolysis pathway, and proteasome complex) underwent ubiquitination in LUSC. Furthermore, 11 subunits of proteasome complex were differentially ubiquitinated in LUSC. These findings demonstrated that ubiquitination was widely present in UPS in LUSC. At the same time, abnormal ubiquitination might affect the functions of the proteasome to promote tumorigenesis and development. This book chapter focused on the status of protein ubiquitination in UPS-related pathways in human LUSC tissues, and provided the scientific data for the elucidation of the specific molecular mechanisms of abnormal ubiquitination during canceration and the development of anti-tumor drugs targeting UPS for lung cancer.

Acknowledgements

The authors acknowledge the financial supports from the Shandong First Medical University Talent Introduction Funds (to X.Z.), the Hunan Provincial Hundred Talent Plan (to X.Z.), and the National Natural Science Foundation of China (Grant No. 81572278 to X.Z.).

Conflict of interest

We declare that we have no financial and personal relationships with other people or organizations.

Author's contributions

X.Z. conceived the concept, designed the book chapter, wrote and critically revised the book chapter, coordinated, and was responsible for the correspondence work and financial support. M.L. designed and wrote the book chapter.

Abbreviations

DUPs	differentially ubiquitinated proteins
E1	ubiquitin-activating enzyme
E2	ubiquitin-conjugating enzyme
E3	ubiquitin ligase
GO	gene ontology
KEGG	kyoto encyclopedia of genes and genomes
LC	liquid chromatography
LUSC	lung squamous cell carcinoma
LUAD	lung adenocarcinoma
MS	mass spectrometry
MS/MS	tandem mass spectrometry
PTM	post-translational modification
PPPM	predictive, preventive and personalized medicine
UPs	ubiquitinated proteins
UPS	ubiquitin-proteasome system

Author details

Xianquan Zhan[1,2,3]* and Miaolong Lu[1,2,3]

1 University Creative Research Initiatives Center, Shandong First Medical University, Jinan, Shandong, China

2 Key Laboratory of Cancer Proteomics of Chinese Ministry of Health, Xiangya Hospital, Central South University, Changsha, Hunan, China

3 State Local Joint Engineering Laboratory for Anticancer Drugs, Xiangya Hospital, Central South University, Changsha, Hunan, China

*Address all correspondence to: yjzhan2011@gmail.com

References

[1] Ciechanover A. The ubiquitin-proteasome proteolytic pathway. Cell. 1994;**79**:13-21

[2] Swatek KN, Komander D. Ubiquitin modifications. Cell Research. 2016;**26**:399-422

[3] Pickart CM, Fushman D. Polyubiquitin chains: Polymeric protein signals. Current Opinion in Chemical Biology. 2004;**8**:610-616

[4] Wilkinson KD. Ubiquitination and deubiquitination: Targeting of proteins for degradation by the proteasome. Seminars in Cell & Developmental Biology. 2000;**11**(3):141-148

[5] Komander D, Rape M. The ubiquitin code. Annual Review of Biochemistry. 2012;**81**:203-229

[6] Kessler BM. Ubiquitin—Omics reveals novel networks and associations with human disease. Current Opinion in Chemical Biology. 2013;**17**:59-65

[7] Adams J, Kauffman M. Development of the proteasome inhibitor Velcade™(Bortezomib). Cancer Investigation. 2004;**22**:304-311

[8] Vij R, Siegel DS, Jagannath S, Jakubowiak AJ, Stewart AK, McDonagh K, et al. An open-label, single-arm, phase 2 study of single-agent carfilzomib in patients with relapsed and/or refractory multiple myeloma who have been previously treated with bortezomib. British Journal of Haematology. 2012;**158**:739-748

[9] Jung SH, Jo JC, Song GY, Ahn SY, Yang DH, Ahn JS, et al. Frontline therapy for newly diagnosed patients with multiple myeloma. Blood Research. 2020;**55**(S1):S37-S42

[10] Jackson GH, Pawlyn C, Cairns DA, Striha A, Collett C, Waterhouse A, et al. Optimising the value of immunomodulatory drugs during induction and maintenance in transplant ineligible patients with newly diagnosed multiple myeloma: Results from myeloma XI, a multicentre, open-label, randomised, phase III trial. British Journal of Haematology. 2020. DOI: 10.1111/bjh.16945

[11] Network CGAR. Comprehensive genomic characterization of squamous cell lung cancers. Nature. 2012;**489**:519

[12] Weiss J, Sos ML, Seidel D, Peifer M, Zander T, Heuckmann JM, et al. Frequent and focal FGFR1 amplification associates with therapeutically tractable FGFR1 dependency in squamous cell lung cancer. Science Translational Medicine. 2010;**2**:62ra93-62ra93

[13] Hammerman PS, Sos ML, Ramos AH, Xu C, Dutt A, Zhou W, et al. Mutations in the DDR2 kinase gene identify a novel therapeutic target in squamous cell lung cancer. Cancer Discovery. 2011;**1**:78-89

[14] Wagner SA, Beli P, Weinert BT, Nielsen ML, Cox J, Mann M, et al. A proteome-wide, quantitative survey of in vivo ubiquitylation sites reveals widespread regulatory roles. Molecular & Cellular Proteomics. 2011;**10**(10):M111.013284

[15] Udeshi ND, Mertins P, Svinkina T, Carr SA. Large-scale identification of ubiquitination sites by mass spectrometry. Nature Protocols. 2013;**8**:1950

[16] Low TY, Magliozzi R, Guardavaccaro D, Heck AJ. Unraveling the ubiquitin-regulated signaling networks by mass spectrometry-based proteomics. Proteomics. 2013;**13**:526-537

[17] Wu Q, Cheng Z, Zhu J, Xu W, Peng X, Chen C, et al. Suberoylanilide

hydroxamic acid treatment reveals crosstalks among proteome, ubiquitylome and acetylome in non-small cell lung cancer A549 cell line. Scientific Reports. 2015;**5**:9520

[18] Qin X, Chen S, Qiu Z, Zhang Y, Qiu F. Proteomic analysis of ubiquitination-associated proteins in a cisplatin-resistant human lung adenocarcinoma cell line. International Journal of Molecular Medicine. 2012;**29**:791-800

[19] Lu M, Chen W, Zhuang W, Zhan X. Label-free quantitative identification of abnormally ubiquitinated proteins as useful biomarkers for human lung squamous cell carcinomas. The EPMA Journal. 2020;**11**:73-94

[20] Sherman BT, Tan Q, Collins JR, Alvord WG, Roayaei J, Stephens R, et al. The DAVID gene functional classification tool: A novel biological module-centric algorithm to functionally analyze large gene lists. Genome Biology. 2007;**8**:R183

[21] Wu J, Mao X, Cai T, Luo J, Wei L. KOBAS server: A web-based platform for automated annotation and pathway identification. Nucleic Acids Research. 2006;**34**:W720-W724

[22] Yuan W-C, Lee Y-R, Huang S-F, Lin Y-M, Chen T-Y, Chung H-C, et al. A Cullin3-KLHL20 ubiquitin ligase-dependent pathway targets PML to potentiate HIF-1 signaling and prostate cancer progression. Cancer Cell. 2011;**20**:214-228

[23] Lim JH, Liu Y, Reineke E, Kao H-Y. Mitogen-activated protein kinase extracellular signal-regulated kinase 2 phosphorylates and promotes Pin1 protein-dependent promyelocytic leukemia protein turnover. The Journal of Biological Chemistry. 2011;**286**:44403-44411

[24] Scialpi F, Malatesta M, Peschiaroli A, Rossi M, Melino G, Bernassola F. Itch self-polyubiquitylation occurs through lysine-63 linkages. Biochemical Pharmacology. 2008;**76**:1515-1521

[25] Pathare GR, Nagy I, Bohn S, Unverdorben P, Hubert A, Körner R, et al. The proteasomal subunit Rpn6 is a molecular clamp holding the core and regulatory subcomplexes together. Proceedings of the National Academy of Sciences. 2012;**109**:149-154

[26] Tanaka K. The proteasome: Overview of structure and functions. Proceedings of the Japan Academy, Series B. 2009;**85**:12-36

[27] Unno M, Mizushima T, Morimoto Y, Tomisugi Y, Tanaka K, Yasuoka N, et al. The structure of the mammalian 20S proteasome at 2.75 Å resolution. Structure. 2002;**10**:609-618

[28] Glickman MH, Rubin DM, Coux O, Wefes I, Pfeifer G, Cjeka Z, et al. A subcomplex of the proteasome regulatory particle required for ubiquitin-conjugate degradation and related to the COP9-signalosome and eIF3. Cell. 1998;**94**:615-623

[29] Etlinger JD, Goldberg AL. A soluble ATP-dependent proteolytic system responsible for the degradation of abnormal proteins in reticulocytes. Proceedings of the National Academy of Sciences. 1977;**74**:54-58

[30] Ermolaeva MA, Dakhovnik A, Schumacher B. Quality control mechanisms in cellular and systemic DNA damage responses. Ageing Research Reviews. 2015;**23**:3-11

[31] Husnjak K, Elsasser S, Zhang N, Chen X, Randles L, Shi Y, et al. Proteasome subunit Rpn13 is a novel ubiquitin receptor. Nature. 2008;**453**:481-488

[32] Deveraux Q, Ustrell V, Pickart C, Rechsteiner M. A 26 S protease subunit

that binds ubiquitin conjugates. The Journal of Biological Chemistry. 1994;**269**:7059-7061

[33] Isasa M, Katz EJ, Kim W, Yugo V, González S, Kirkpatrick DS, et al. Monoubiquitination of RPN10 regulates substrate recruitment to the proteasome. Molecular Cell. 2010;**38**:733-745

[34] Lipinszki Zn KL, Deák Pt UA. Ubiquitylation of drosophila p54/Rpn10/S5a regulates its interaction with the UBA–UBL polyubiquitin receptors. Biochemistry. 2012;**51**:2461-2470

[35] Cohen-Kaplan V, Livneh I, Avni N, Fabre B, Ziv T, Kwon YT, et al. p62-and ubiquitin-dependent stress-induced autophagy of the mammalian 26S proteasome. Proceedings of the National Academy of Sciences. 2016;**113**: E7490-E7499

[36] Vilchez D, Boyer L, Morantte I, Lutz M, Merkwirth C, Joyce D, et al. Increased proteasome activity in human embryonic stem cells is regulated by PSMD11. Nature. 2012;**489**:304-308

[37] Van Damme P, Lasa M, Polevoda B, Gazquez C, Elosegui-Artola A, Kim DS, et al. N-terminal acetylome analyses and functional insights of the N-terminal acetyltransferase NatB. Proceedings of the National Academy of Sciences. 2012;**109**:12449-12454

[38] Wang X, Chen C-F, Baker PR, P-l C, Kaiser P, Huang L. Mass spectrometric characterization of the affinity-purified human 26S proteasome complex. Biochemistry. 2007;**46**:3553-3565

[39] Hendriks IA, Treffers LW, Verlaan-de Vries M, Olsen JV, Vertegaal AC. SUMO-2 orchestrates chromatin modifiers in response to DNA damage. Cell Reports. 2015;**10**:1778-1791